景观设计学教育参考丛书

沙漠化地区可持续景观与城市设计
Sustainable Landscape and Urban Design in Desertification Region

北京大学与瑞典隆德大学合作设计研讨课成果
A Joint Workshop Report of the Peking University and Lund University, Sweden

主编 韩西丽 ［瑞典］彼得·斯约斯特洛姆 ［瑞典］马丁·阿尔福尔克
Edited by Xili Han, Peter Siöström and Martin Arfalk

中国建筑工业出版社
CHINA ARCHITECTURE & BUILDING PRESS

图书在版编目（CIP）数据

沙漠化地区可持续景观与城市设计/韩西丽，[瑞典]斯约斯特洛姆，[瑞典]阿尔福尔克主编．—北京：中国建筑工业出版社，2016.7
（景观设计学教育参考丛书）
ISBN 978-7-112-19320-2

I.①沙… II.①韩…②斯…③阿… III.①干旱区—城市景观—景观设计—研究—西北地区 IV.① TU-856

中国版本图书馆CIP数据核字（2016）第067096号

责任编辑：杜　洁　李　杰
责任校对：刘　钰　张　颖

景观设计学教育参考丛书
沙漠化地区可持续景观与城市设计
主编　韩西丽　［瑞典］彼得·斯约斯特洛姆　［瑞典］马丁·阿尔福尔克

*

中国建筑工业出版社出版、发行（北京西郊百万庄）
各地新华书店、建筑书店经销
北京京点图文设计有限公司制版
北京盛通印刷股份有限公司印刷

*

开本：787×1092毫米　1/16　印张：7¼　字数：180千字
2016年7月第一版　2016年7月第一次印刷
定价：68.00元
ISBN 978-7-112-19320-2
（28529）

版权所有　翻印必究
如有印装质量问题，可寄本社退换
（邮政编码 100037）

"景观设计学教育参考丛书"总序

景观设计学是对于土地及土地上空间和物体所构成的地域综合体的分析、规划、设计、改造、管理、保护和恢复的科学和艺术。景观设计学尤其强调对于土地的监护与设计,是一门建立在广泛的自然科学和社会科学基础上的综合性较强的应用学科,与建筑学、城市规划、环境艺术等学科有着紧密的联系,并需要地理学、生态学、环境学、社会学等诸多学科背景的支持。

在我国城市迅速发展的背景下,景观设计学所承担的责任显得愈发重要。在城市建设快速发展的情况下,在前所未有的发展机遇面前,我国同样面临着严峻的挑战。由于长期以来片面追求经济发展,我国显现出日益突出的人地关系危机。值得庆幸的是,近些年来政府管理者清醒地认识到这些问题,及时做出转变,明确提出用科学发展观指导城市建设,强调人与自然和谐共存的可持续发展理念,中共十七大更加明确提出生态文明的重要性。在这样宏观政策的指引下,面对时代赋予的使命,我国的景观设计专业人才培养也显得愈发重要,培养适应我国当前需要的景观设计专业人才已刻不容缓。

然而,总体来说,我国当代的景观设计学教育还处在初级阶段,学科建设与教学体系还很不完善,各学校之间各自独立,没有形成相对统一的教学模式与教育体系。这对于我国景观设计学学科发展和人才培养显然是不利的。

面对如此的趋势与需求,以北京大学为首的各高等院校相继开设景观设计学专业,学科教育联盟雏形已现,教学体系也在探索中逐步走向完善。在各高等院校大力支持与配合下,北京大学建筑与景观设计学院在吸取国外学科建设模式经验的基础上,逐步探索出一套适应于我国国情的景观设计学专业与学科教育体系。为了促进我国景观设计学科发展,为国家培养和输送更多的专业人才,北京大学景观设计学研究院牵头联合各院校推出景观设计学教育参考丛书。本套丛书收录了优秀的景观设计学课程教学案例,旨在为我国景观设计学专业教育提供更新、更完善的思路,为开展相关专业的各院校搭建一个交流平台,使学科得以良好健康地发展,为我国构建可持续发展的和谐人地关系贡献更多专业人才。

俞孔坚

课程成员

北京大学教师

韩西丽 副教授
北京大学深圳研究生院城市规划与设计学院
北京大学建筑与景观设计学院

北京大学学生

卜小
曹安康
陈曦
程红济
范非
韩冰
衡先培
胡岳
黄俊博
贾琳
蒋晓峰
寇淼
李绪文
林浩文
刘丽春
刘芹芹
刘玥
马静薇
那然
倪冰
汤博深
王朝倩
王烨
吴梦霞
吴欣玥
向林森
谢芳丽
杨雪
余洁燕
袁远
张冰洁
张仁达
张桐伟
张玥一
赵茜

Members

Peking University Faculty

Xili Han, Associate Professor
School of Urban Planning and Design, Shenzhen Graduate School
College of Architecture and Landscape Architecture

Peking University Students

Bu Xiao
Cao Ankang
Chen Xi
Cheng Hongji
Fan Fei
Han Bing
Heng Xianpei
Hu Yue
Huang Junbo
Jia Lin
Jiang Xiaofeng
Kou Miao
Li Xuwen
Lin Haowen
Liu lichun
Liu Qinqin
Liu Yue
Ma Jingwei
Na Ran
Ni Bing
Tang Boshen
Wang Zhaoqian
Wang Ye
Wu Mengxia
Wu Xinyue
Xiang Linsen
Xie Fangli
Yang Xue
Yu Jieyan
Yuan Yuan
Zhang Bingjie
Zhang Renda
Zhang Tongwei
Zhang Yueyi
Zhao Qian

瑞典隆德大学教师

Lars-Henrik Ståhl,
瑞典隆德大学建筑与环境系主任、理论和应用美学教授
Peter Siöström,
瑞典隆德大学可持续城市设计项目主管、副教授、瑞典建筑师协会注册建筑师
Martin Arfalk,
瑞典隆德大学可持续城市动力课程主任、瑞典Mandaworks, 设计公司主管、景观建筑设计师
Nicholas Bigelow,
瑞典隆德大学可持续城市动力课程助理、瑞典Mandaworks, 设计公司城市设计师
张铁林,
瑞典隆德大学水资源工程系副教授
张楚晗,
瑞典隆德大学可持续城市动力课程助理、城市设计师

Lund University Faculty

Mr. Lars-Henrik Ståhl, Professor in Theorteical and Applied Aesthetics Head of Department, Dept. of Architecture and Built Environment Lund University, Sweden
Mr. Peter Siöström, Ass. professor, Architect SAR/MSA Director of SUDes, Chairman of Ax:son Johnson Institute of Sustainable Urban Design at Lund University, Sweden
Mr. Martin Arfalk, Landscape Architect Director of Mandaworks Course Leader Urban Dynamics, SUDes at Lund University, Sweden
Mr. Nicholas Bigelow, Urban Designer Mandaworks AB Assistant Course Leader Urban Dynamics, SUDes at Lund University, Sweden
Mr. Linus Tie Zhang, Ass. Professor in Water Resources Department of Water Resouces Engineering, Lund University, Sweden
Ms. Chuhan Zhang, Graduate Urban Designer Teaching Assistant and Trip Coordinator Urban Dynamics, Lund University, Sweden

瑞典隆德大学学生

Marino Stefánsson
Gisli Gudmundsson
Felix Perasso
Alex Seltea
Claudia Raquel Bessa e Meneses Serra
Miriam Castel Cierco
Karin Andersson
Kajsa Henriksson
Elina Berezovka
Anna Kravec
Sybille de Cussy
Suzanna Rubino
Payam Faalzadeh
Piotr Decko
Mårten Espmarker
Andreas Ask
Katarina Vondrova
Hanne Bendixen
Andreas Mayor
Olof Eriksson
Young Ill Kim
Jaime E
Jenny Nygren
Katarina Hansson
Alina Velaviciute
Eirini Oikonomopoulou
Cyril Pavlu
Constantin Milea
Gabriella Klint

Lund University Students

Marino Stefánsson
Gisli Gudmundsson
Felix Perasso
Alex Seltea
Claudia Raquel Bessa e Meneses Serra
Miriam Castel Cierco
Karin Andersson
Kajsa Henriksson
Elina Berezovka
Anna Kravec
Sybille de Cussy
Suzanna Rubino
Payam Faalzadeh
Piotr Decko
Mårten Espmarker
Andreas Ask
Katarina Vondrova
Hanne Bendixen
Andreas Mayor
Olof Eriksson
Young Ill Kim
Jaime E
Jenny Nygren
Katarina Hansson
Alina Velaviciute
Eirini Oikonomopoulou
Cyril Pavlu
Constantin Milea
Gabriella Klint

目 录

10	**研讨课程简介**
16	**设计方案**
20	城市探测器——城市生态基础设施规划设计
29	看不见的水——城市生态基础设施规划设计
40	武威市郊湿地公园规划设计——城郊低洼滞水区的改造与利用
50	N 次方公园——城市公园改造提升
60	土城——历史街区保护与更新
67	流动市场——历史街区保护与更新
76	城市动脉——城市扩展区可持续城市设计
82	城市绿洲——城市扩展区可持续城市设计
88	挑战荒漠——城市扩展区可持续城市设计
95	河流之上——城市扩展区可持续城市设计
101	包容的武威——城市扩展区可持续城市设计
107	城市群岛——城市扩展区可持续城市设计
115	**现场踏勘**
116	**致谢**

Contents

10	**Workshop Introduction**
16	**Design Proposals**
20	City Hunter——Urban Ecological Infrastructure Planning and Design
29	Invisible Water——Urban Ecological Infrastructure Planning and Design
40	Design of Wetland Park in Suburb of Wuwei——Transformation and Utilization of Suburban Informal Low-lying Water Storage Areas
50	N-th Power Park——Upgrading of the Existing City Park
60	Soil City——Protection and Renewal of the Historic District
67	Moving Market——Protection and Renewal of the Historic District
76	Urban Arteries——Sustainable Urban Design of City Expansion Area
82	Urban Oasis——Sustainable Urban Design of City Expansion Area
88	Challenging Desert——Sustainable Urban Design of City Expansion Area
95	Over the River—— Sustainable Urban Design of City Expansion Area
101	Inclusive Wuwei—— Sustainable Urban Design of City Expansion Area
107	The Urban Archipelago——Sustainable Urban Design of City Expansion Area
115	**Site Visit**
116	**Acknowledgement**

Workshop Introduction
研讨课程简介

随着中国经济的飞速发展，中国经历了史无前例的快速城市化进程，国家统计局数据显示中国城市化率在2014年已经达到54.77%。随着发展的不断深化，诸如环境污染、水资源短缺、土地荒漠化、城乡差距加大、文化丧失等问题日益严峻，城镇建设迫切需要找到一条可持续发展的途径。

我国西北干旱区生态环境脆弱，气候干旱，降水稀少，蒸发量大，水资源极为短缺。武威市位于戈壁滩的边缘，具有干旱缺水的气候特征，年降雨量低于100mm，武威的干旱气候对城市发展造成严重的约束。加之近年来生态环境的变化，工业化和城市化进程的加快，导致水资源供需的严重失衡，武威市人均占有水资源量是全省平均水平的1/2和全国水平的1/4，水资源利用率低下，用水分配不平衡，农业灌溉用水占所有用水量80%以上，结构性缺水严重。长期的地下水开采，导致地下水位不断下降，土地荒漠化日趋严重。石羊河下游的民勤绿洲处在巴丹吉林沙漠和腾格里沙漠之间，是阻断两大沙漠汇合的重要屏障，随着本地区水资源问题日益加剧，土地荒漠化态势愈发严重，如不转变发展模式，民勤将有可能成为第二个罗布泊，不仅当地人民将沦为生态难民，更将威胁全国的生态安全。面对严峻的环境挑战，武威迫切需要进行水敏感性景观与城市设计。武威市规划在2030年城市人口规模增加两倍达到80万，这一发展规模为将武威发展成为中国水及能源明智城市的典范提供了良机。本次国际设计研讨课程，希望通过对以上问题的探讨和研究，为干旱缺水地区寻求一种更为可持续的城市建设之路。

研讨目标：
（1）将景观设计学与城市设计专业知识相结合，提出武威市可持续发展的愿景；
（2）学生的设计方案需要呈现出水敏感城市设计的解决策略，旨在为武威这样的干旱气候环境下的城市提出其可持续发展的城市设计模式；
（3）本次研讨鼓励学生们大胆想象，将所观察到的场地发展潜力尽可能放大，分析场地的制约因素，试图预想出水及能源使用更加明智以及更加可持续的未来。

研讨主题：
（1）"多尺度城市生态基础设施规划与设计"

旨在了解和学习如何规划城市生态基础设施，并使其作为未来城市生存的绿色基底，同时引导城市的发展方向，学生们被要求从宏观上思考城市增长策略、水资源管理策略、绿地结构及其他城市基础服务设施。

Due to the rapid development of economy, China has been experiencing the rapid and unprecedented process of urbanization. Statistics show that the urban population share in China rose significantly to 54.77% in 2014. As development continues, problems such as environmental pollution, water shortage, land desertification, growing gap between urban and rural standards of living, and erasure of authentic local cultures have become more and more serious. Finding a pathway to sustainable development has never been more urgent.

The ecological environment in the north-western arid area of china is very fragile, the major features of this area are less rainfall, great evaporation, and hence it is short of water resources. Wuwei, as a typical oasis town of Hexi corridor, the process of industrialization and urbanization has generated severe imbalance of supply and demand of water resource. The water resource occupied per capita in Wuwei is only 1/2 of the province average level,1/4 of the China. The overdrawn exploitation of groundwater has led to the water table down, desertification has become increasingly serious. In the face of these environmental challenges, Wuwei is planning to double the city's size by 2030 to around 800,000 people. The scale of this development will give Wuwei a unique opportunity to present a new model for a water-smart city in China. This international design seminar expects to engage squarely with a number of these issues in a search for a more sustainable future.

Goal with workshop:

(1) Merge knowledge of the landscape with an urban design approach to make a vision.

(2) Students' proposals will present water sensitive urban design solutions and aim to develop urban design models for Wuwei's arid climate.

(3) The students are encouraged to dream big and strive to maximize observed potentials, test the limits of the site's context, and attempt to envision a more water-smart and sustainable future.

Design Topics:

(1) Multi-scale ecological infrastructure planning and design

The aim is to understand how to plan regional ecological infrastructure first, and make it work as the green basis of the future city and guide urban growth as well. Students were asked to give an overall vision for the water-smart growth of Wuwei and to include a strategy for urban growth, water management, green space, and other supporting infrastructure.

需要外来水源维持的青土湖湿地　Qingtuhu Wetland Sustained by External Water

基础设施落后　Laggard Infrastructure

（2）"城郊低洼滞水区的改造与利用"

该湿地区域位于武威市北部边缘处，该组的研讨目标是去解决郊区城市化过程中出现的问题以及城市边缘带修建大型公园的服务效率低下问题。另外，在保护城郊湿地的同时，通过将分散的低洼滞水区相互连接和疏通，提高城市的洪水消解能力，并保护和利用城市稀缺的淡水资源。

（3）"历史街区保护与更新"

旨在了解在保护历史街区的同时，学习如何将其转变成宜居的、富有吸引力的城市街区，赋予其新的生命。从邻里关系、公共空间结构、乡土建筑材料、街区小气候、传统与现代生活方式融合途径等角度对历史街区进行思考，并进行更新设计。

（4）"城市公园改造提升"

目标在于找到溶解公园的途径，将公园的功能多样化，凸显公园的生态与社会服务特色，寻找公园与其周边的街区协同发展的途径，进而提高现状城市公园的景观绩效。

（5）"城市扩展区可持续城市设计"

紧紧围绕武威水敏感城市设计这一核心概念，考虑旧城中心与新的高铁站之间的联系。从场地实际问题、挑战与机遇出发，发展出自己的设计概念，提出城市扩展区的建设策略及水敏感性城市设计策略。在生态上，在街区尺度上重新建立了良性的水文循环系统，把中水回用设施融入城市社区，融入市民生活；在文化上，充分挖掘武威本土文化特质，力图在城市建设中延续当地文脉；在社会生活上，仔细聆听当地人的故事，在设计上回应人们对场所精神以及对人性化生活的呼唤。

总体来讲，本次设计研讨课程将目光投向干旱荒漠化地区的城市建设问题，用生态的理念探索武威市可持续发展的可能途径，提出了许多富有创见的设计方案，也丰富了景观与城市设计实践。

(2) Transformation and utilization of suburban informal low-lying water storage areas

This wetland area is located on the north suburban area of Wuwei. This topic aims to solve the problem out of the suburban area's urbanization and the problem of low efficiency of big park built on the edge of cities. In addition, beside the urban wetland protection, try to connect the dispersed low-lying water storage areas, improve the urban flood digestion ability, and protect and utilize the scarce fresh water resources.

(3) Protection and renewal of the historic district

Another goal is to learn how to protect the historical apartment blocks while transforming them to bring new life in a more livable and attractive built environment.Think and put forward renewal proposal of the historical district from neighborhood relationship, structure of public open spaces, local construction material, microclimate at site scale, and the integration of traditional and modern lifestyle.

(4) Upgrading of the existing city park

Finally we shall figure out a way to dissolve park boundaries and bring multiple functional dimensions to it, make it work together with the development of the surrounding community. Improve the landscape performance of city park.

(5) Sustainable urban design of the urban expansion area

Students were asked to propose a project to articulate the historic city center, the design site, and the planned high-speed rail station. Centered on the core concept of water-smart design, the students developed their own projects based on the actual issues, challenges and opportunities, and proposed their own water-smart urban development strategies. To foster ecological construction projects included initiatives to repair the water system, design water recycling facilities to blend with the urban landscape and with citizen life. Cultural construction gave full play to the facilities and institutions required to sustain urban culture. To enrich the local social life, they listened carefully to local people's stories and responded to the call for a greater sense place-based belonging, spirit and life-style.

In general, this workshop focused on urban adaptation to drought in a naturally arid region, using ecological concepts to explore the potential pathways to the sustainable development of Wuwei, developing visionary proposals to enrich the practice of the landscape architecture.

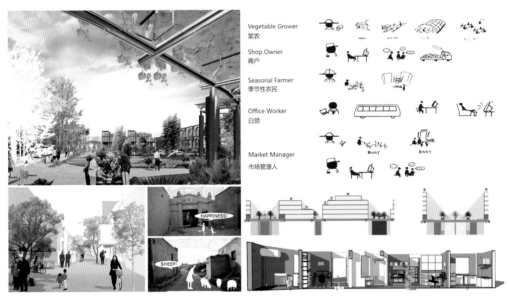

两校学生合作部分成果　Part of Outcome Accomplished by Students from the Two Universities

北京大学最终汇报　Final Presentation in Peking University

Design Proposals
设计方案

城市生态基础设施规划设计
Urban Ecological Infrastructure Planning and Design

城市探测器——城市生态基础设施规划设计
City Hunter——Urban Ecological Infrastructure Planning and Design

林浩文、卜小、王朝倩、那然、余洁燕
Lin Haowen, Bu Xiao, Wang Zhaoqian, Na Ran, Yu Jieyan

当前武威市农业灌溉消耗80%的总用水量,水资源分配不均,导致出现地下水位急剧下降、荒漠化严重等生态问题。该方案依据地下水的敏感度选取当地四种指示植被的种植来引导城市不同模式的发展,从根本上减少地下水的过度开采。从指示试点林带培育、指示植物景观建设、指示植物能源开发三个阶段引导城市可持续发展,构建武威市生态基础设施。

Today agricultural irrigation of Wuwei consumes 80% of the total water consumption. The great unbalance of water resource leads to many ecological problems such as the falling groundwater table and desertification. This project used four local plants to guide the different city development modes based on the sensitivity to the groundwater resource and its conservation. The indicator plants go through three phases, from experimental forest to plant landscape, then finally become a part of energy sources of the city. They promote sustainable city development and build an ecological infrastructure for Wuwei.

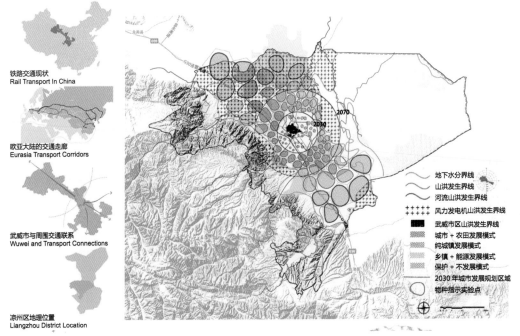

区位分析
Location Analysis

宏观区域景观战略图 Regional Landscape Strategy Map

宏观分析
Macroscopic Analysis

武威市凉州区位于中国甘肃省中部，凉州区内农业灌溉消耗80%的总用水量，造成河道干涸，地下水位下降，进而造成植被根系缺水，水土流失，在北部腾格里沙漠、河西走廊管道式地势荒漠化逐渐严重，沙漠不断向城市蔓延。

宏观层面主要通过减少农耕用地、指导城市有序发展和转变产业结构三个方向进行设计。指示植物为各场地寻找到适宜的发展模式提供了可能。根据指示植物未来的长势，将凉州区分为农田城市、纯城市、能源产业、水土涵养四种发展模式，并划定2030年第一期、2070年第二期的发展范围。

利用当地四种植被对地下水的敏感度（敏感度：白刺 *Nitraria tangutorum* > 梭梭 *Haloxylon ammodendron* > 沙拐枣 *Calligonum arborescens* > 沙枣 *Elaeagnus angustifolia*）指示城市的不同发展模式，不同的地下水水位支持相适宜的发展模式，从根本上减少地下水的过度开采。成片的指示植物在城市未发展地区防风固沙阻止荒漠化。城市发展至指示试点时，部分指示植物融入城市，成为城市景观，剩余指示植物用以生产能源支持城市发展。指示植物经历指示试点林带培育、指示植物景观建设、指示植物能源开发三个阶段，促进城市可持续发展，构建武威市生态基础设施。

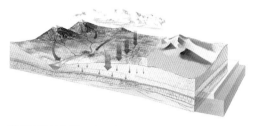

凉州区总体剖面　The Section of Liangzhou District

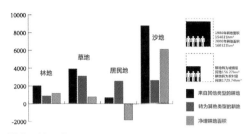

基本用地现状　Land Condition

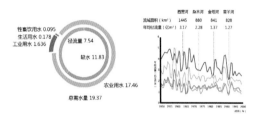

基本用水现状　Water Condition

Liang zhou district of WuWei city is located in the middle of GanSu province, China.In LiangZhou district agricultural irrigation consumes 80% of the total water consumption, causing the river dried up, the underground water levels fall. What's more, vegetation root system lacks water, soil and water loses, and under the joint action of the Tengery desert in the northern, hexi corridor pipeline terrain and northwest monsoons Liangzhou district gradually suffers serious desertification. Desert continuously spread to the cities.

At macro level, the city design is mainly carried on by reducing the agricultural land and guiding orderly urban development and transforming industrial structure. Indicator plants provide possibilities to find suitable development model for the site. According to the indicator plants growing situation in the future, LiangZhou will be divided into four kinds of development modes, farmland city mode, pure city mode, energy industry mode and soil and water conservation mode, and draw a phased development stage (i,2030ii,2070)

Using the four local plants (*Nitraria tangutorum*, *Haloxylon ammodendronto*, *Calligonum arborescens*, and *Elaeagnus angustifolia*) we shall guide the different city development modes based on the sensitivity to the groundwater availability, and help reduce the excessive exploitation of groundwater sources. (*Nitraria tangutorum*>*Haloxylon ammodendron*>*Calligonum arborescens*> *Elaeagnus angustifolia*).The indicator plants in the experimental unit will be used to prevent the desertification, some of which will blend in the city landscape. The remaining indicator plants will product the energy to sustain the whole city. Indicator plants go through three phases, from experimental forest to plant landscape, then finally become a part of energy sources of the city. They promote sustainable city development and build an ecological infrastructure for Wuwei.

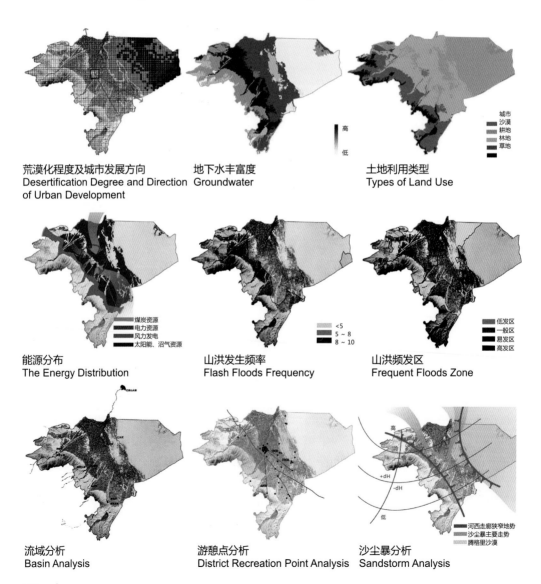

荒漠化程度及城市发展方向
Desertification Degree and Direction of Urban Development

地下水丰富度
Groundwater

土地利用类型
Types of Land Use

能源分布
The Energy Distribution

山洪发生频率
Flash Floods Frequency

山洪频发区
Frequent Floods Zone

流域分析
Basin Analysis

游憩点分析
District Recreation Point Analysis

沙尘暴分析
Sandstorm Analysis

中观规划
Mesoscopic Plan

在原有的生态基础设施的基础上构建防风固土林带、沿河生态廊道与地下水科普廊道三条生态廊道，并结合食品安全格局、能源安全格局与南部水源安全格局三种格局，形成2030年一期景观生态安全格局。基于景观安全格局的重叠频率，敏感性高的地区位于场地北部。我们选取老城与三条河流交汇的冲积扇之间的地区作为中观的规划范围，在生态基础设施的服务半径内设置主要公交站并延续老城街道形式形成新的道路网格。指示植物试点区域之间通过基础设施和交通系统连接逐渐形成成熟的城市，指示植物融入城市，与旧城、新城、能源农田、生态河道相结合，形成城市景观。

On the basis of the original ecological infrastructure, three ecological corridors, windproof forest corridor, ecological corridor along the river and groundwater corridor for education, are proposed, and with the combination of food, energy and water security patterned.Proposed landscape ecological security pattern will be finished by 2030. The service radius of the ecological infrastructure includes the main bus station and road grid, the old city, the newly designed city, energy fields, and ecological river course within the city landscape as a whole. Experimental units of indicator plants are linked through public infrastructure and traffic system, then gradually grow into a mature urban city.As city develops, indicator plants are blended into the city.

河流景观效果图　Perspective of Waterfront Landscape

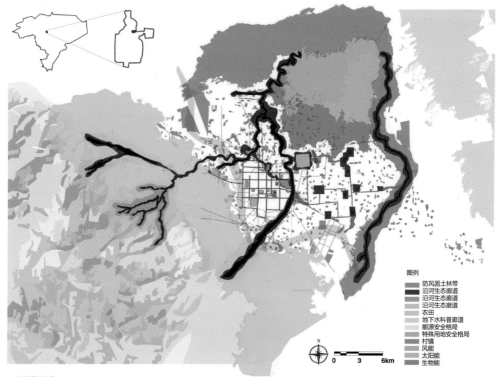

景观生态安全格局　Ecological Security Patterns

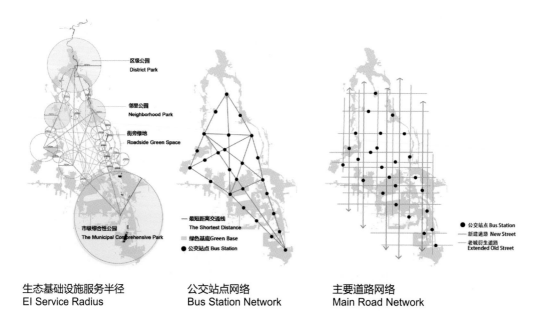

生态基础设施服务半径
EI Service Radius

公交站点网络
Bus Station Network

主要道路网络
Main Road Network

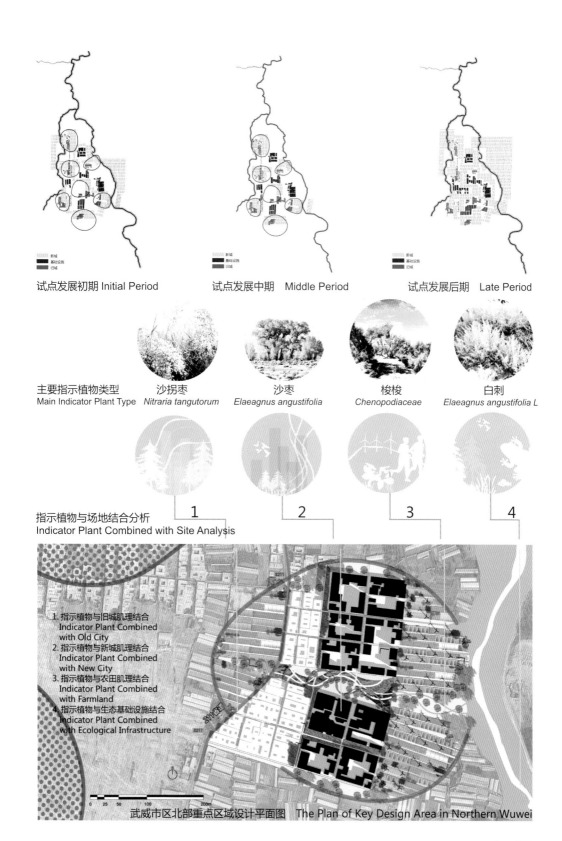

总剖面 A-A　Master Section A-A

街道景观剖面 B-B
The Section of Street B-B

农田景观剖面 C-C　The Section of Landscape in Farmland C-C

河流景观剖面 D-D
The Section of Landscape Along River D-D

科普廊道效果图
Effect Drawing of Popularization of Science Corridor

农田景观效果图　Effect Drawing of Landscape in Farmland

微观设计
Microcosmic Design

此处场地是中观设计场地中代表性的一块,包含了新设计的建筑院落、街旁绿地和农田的一部分。利用当地的黄土和植物纤维作为建筑外表皮,与当地保留老建筑和谐统一,建筑中加入廊道相连,构造舒适的院落尺度。院落中央以下沉的广场作为主要活动场地,营造适宜活动的丰富空间体验。

街旁绿地主要有一条自行车道和一条人行道,道路中间以地形间隔,高处种植梭梭、白刺、樟子松、小叶杨、女贞、千头柏形成植物群落,这些植物具有耐干旱、耐沙土、耐阴、耐寒、滞尘能力强等特点,低洼地形用于雨水的暂时储存,有利于涵养地下水。农田中以梭梭间隔粮食作物的形式种植,辅以风车,充分利用当地的风能和植物能,体现可持续的弹性设计思想。

The site contains a new designed yard, part of the street green space and farmland. Local loess-based soils and plant fiber are used outer skin of buildings articulated by corridors with sunken squares in the center of the courtyard as the main activity space. These features will all make a contribution to an appropriate activity space with the potential for enriched experience.

Roadside green space has a bicycle lane and a sidewalk, while the middle of the road is spaced by terrain, where plant *Haloxylon ammodendron*, *Nitraria tangutorum*, *Pinus sylvestris*, *Populus simonii*, *Ligustrum lucidum*, *Platycladus* and so on. These plants are hardy and resistant to cold, drought, and shade. They are tolerant of rainwater storage on low-lying land which will help to conserve groundwater. Crops are interplanted with *Haloxylon ammodendron* in the farmland, where the windmills make full use of the local wind and plants consistent with sustainability objectives.

场地设计平面图 Master Plan

人视效果图 Perspective

屋顶雨水灌溉　　当地建筑材料　　乡土植物群落
Rainfall Irrigation　　Local Materials　　Local Plants

剖面图 D—D' Section D—D'　　剖面图 E—E' Section E—E'

看不见的水——城市生态基础设施规划设计
Invisible Water——Urban Ecological Infrastructure Planning and Design

黄俊博、张桐伟、吴欣玥、范非、陈曦
Huang Junbo, Zhang Tongwei, Wu Xinyue, Fan Fei, Chen Xi

当前武威市面临的最大问题是水资源分配不平衡，出现了荒漠化等生态问题。本方案通过改变城市水资源利用结构来平衡地下水位，以改变地表景观，在解决生态问题的同时为市民创造更多的绿色空间。

The current crisis of Wuwei is the unbalance of water resource which results in many ecological problems such as desertification. This project aims to balance the groundwater through changing the water resource utilization structure.It can creat more green spaces for citizens while relieving ecological crisis.

宏观规划
Macroscopic Plan

遗产点分析：凉州区大量人文遗产主要集中在西北部。自然资源主要集中在南部山区。

水文分析：境内共有七座水库，四条河流。地下水流向由南至北，盆地地下水水质恶化，矿化度由南至北升高，到民勤地区已不能饮用和灌溉农田。

荒漠化分析：南部水土流失严重；中部绿洲地区大量消耗水资源，地下水位下降；北部地区大量抽取地下水灌溉农田，造成农田中矿物富集，农田盐碱化。

粮食安全分析：凉州区共有六大农业灌区，各有独立的灌溉系统；有四大粮食生产基地。由于地下水位下降，农业生产效率越来越低，无法实现自给自足。

清洁能源分析：北部处于河西走廊通风口，风速大且稳定，有大量闲置土地，适合发展风能；东部沙漠地区适合发展太阳能；农村地区生产的作物秸秆发展沼气能源。

坡度分析：场地南部坡度较大，土壤结构疏松，易发生水土流失；上游农田开垦，河水含沙量增加，泥沙沉积在下游，抬升河床。

遗产点分析
Heritage Analysis

水文分析
Hydrological Analysis

坡度分析
Gradient Analysis

粮食供应分析
Food Support Analysis

清洁能源分析
Sustainable Energy Analysis

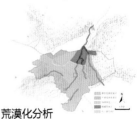
荒漠化分析
Desert Analysis

Heritage Analysis: Cultural heritage is oriented mainly towards the northwest of Liangzhou District, while natural heritage is in the south.

Water Analysis: There are 7 reservoirs, 4 rivers. Underground water flows from south to north. Mineralization rises from south to north, and in Minqin, the water can not be used for drinking and irrigating.

Desert Analysis: In the south, water loss and soil erosion are substantial while people use much water in the middle oasis area that the level of ground water is declining.

Food Safety Analysis: There are 6 agricultural irrigated areas in Liangzhou district, which have separated irrigation system. Also there are 4 food production bases.

Clean Energy Analysis: The north area is in the Hexi Corridor, where wind-speed is high and stable. Thus it is suitable for capturing wind energy. The east desert area is well-suited for harvesting solar energy.

Slope Analysis: The slope in the south site is steep. The soil is loose. Water loss and soil erosion happen easily. Much reclamation of farmland in upstream causes more sand in river. The silt sediment in downstream raises river bed.

浅山区存在地质断层，是地表水补给地下水；市区周围地下水位下降，地下水失去了对地表水的补给能力，是地表水补给地下水；市区北部细土平原地下水位高，在非汛期对石羊河有补给能力，是地下水补给地表水，汛期是地表水补给地下水；红崖山水库至民勤盆地的地下水完全来自于地表水的补给。

区域南部作为水土涵养区域限制农田开垦，保障地下水补给过程的安全。中部沿跨境铁路构建游憩廊道，加强区域南北联系；西北部沿现存绿带构建三条防风林廊道，遏制巴丹吉林沙漠向南推进，减小沙尘暴对城市的影响；北部种植盐生植物，设置盐碱地生态治理区；东部构建粮食运输廊道和能源廊道，保障城市的粮食和能源供给。沿四条河流构建河流防护廊道，遏制水土流失；规划的总体战略是：南部涵养水土，中部塑造绿洲，北部防治风沙。

There are geological faults in mountainous areas, where surface water recharges underground aquifers. Groundwater levels decline near urban districts and are insufficient to renew surface water resources. Fine soil plan is in the north of urban area, and there the level of groundwater is high enough to replenish surface water. In dry season, it can provide water to Shiyang River, and in flood season, the orientation of water supply is reversed. The surface water supplies underground water entirely in the area between Hongyashan Reservoir and Minqin Basin.

The south of the site is proposed as water and soil conservation district, and development of farmland is limited to assure the safety of underground water replenishment. In the middle, recreation corridors are planned along the cross-border railroad, to enhance the connection of south and north districts.

In northwest, three corridors of windbreak woods are designed to prevent the southward expanding of Badain Jaran Desert and to reduce the influence of sand storm in the city. In the northern city, set as saline-alkali soil ecological management area, halophyte will be planted. Food transportation corridor and energy corridor are in the east to assure the supplementation of food and energy in the city. Along the four rivers, we plan to build 4 river-protection corridor, to prevent water lose and soil erosion. The main strategy of the plan is like this: soil and water conservation in the south, oasis improvement in the middle, and sand prevention in the north.

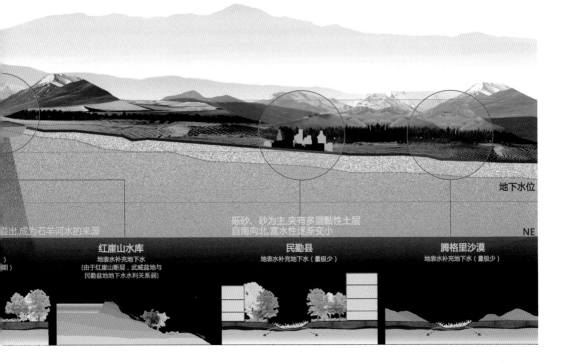

在城区南部河段两岸设置10m左右的缓冲区，并在缓冲区内设置步行道，防止水土流失的同时，也为人创造了亲水空间；在城区内部，将原有的硬质河道恢复为自然衬底，变窄河道，并设置锯齿状亲水平台，沉积上游泥沙；在城区北部河段种植盐生植物，作为生态恢复示范区，并在河道两岸设置20m左右的缓冲区。

On both sides of the river in the southern city, 10m buffer would be set in which there are walkways. The buffer can prevent soil erosion, and can also create hydrophilic space; In the urban area, the current hard straight river bank would be restored to natural serrated bank to accumulate sediment deposition from upstream river; In the northern city, ecological restoration demonstration zone, about 20m buffer would be set on both sides in which halophytes grow.

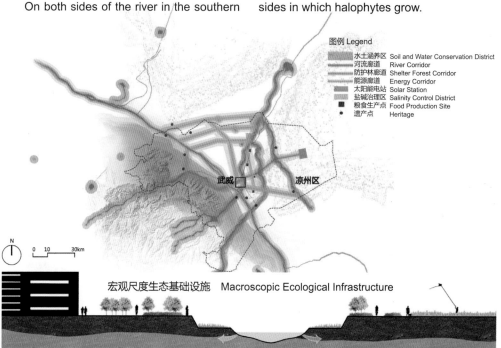

宏观尺度生态基础设施　Macroscopic Ecological Infrastructure

南部城市边缘　South Urban Fringe

城市内部　Downtown

细土平原　North Area(Salt-alkali Soil)

中观规划
Mesoscopic Plan

现存的公共绿地分散在城区周边,利用这些公共绿地作为中观结构的面元素;在城市外围修建自行车道,结合城市中心的机动车道和东部河流,构建一个覆盖整个城区的完整连续的自行车道廊道,作为中观结构的线元素;机动车道将城市分割为若干社区,在每个社区中选取地势较高处修建灰水处理池,处理的生活污水用于灌溉自行车道两侧绿带,并补充地下水,同时结合城市中的遗产点,共同作为中观结构的点元素。通过"点-线-面"结构,达到降低城市能源消耗、减少城市生活污水排放、补充城区地下水、增加城市绿地、改善城市景观的目的。

The existing public green spaces are dispersed near the urban area. Bicycle paths are planned for the external area around the city. Combined with the central roadway and the river in the east, a complete and continuous network of bicycle pathways integrate the whole urban area. Communities are divided and defined by roadways. In every community, there is a gray water purification pond for domestic waste water, which is built at a higher elevation to irrigate the green belt along both sides of the bicycle path, and to recharge ground water. The "point-line-surface" framework is planned to reduce energy and sewage discharge, supply ground water, increase green space, and add interest and variety to the urban landscape.

根据武威市区3万人/km²人口密度,每人每天产生80L生活灰水,每平方米处理池的日处理能力为2.5m³。在城区共设置20~30个处理池,每个处理池的面积在200m²左右,每个处理池的服务半径在200m(2000~6000人)左右。这些处理池在处理生活灰水的同时,也能达到增加户外空间,调节小气候,改善景观的作用。

The population density of Wuwei city is 30000 / km². 80L life grey water would be produced per person per day, and daily processing capacity of soil filter basin per square meter is 2.5 m³. 20 ~ 30 ponds are set in the urban area, each filter basin area is around 200 m², and each basin's service radius is about 200m(for 2000 ~ 6000people). The ponds can increase open space, adjust microclimate while disposing the grey water.

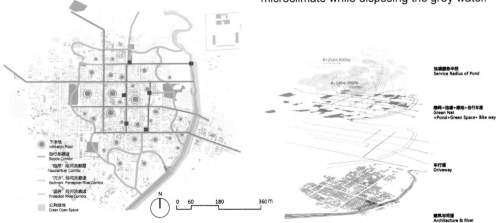

中观尺度生态基础设施
Mesoscopic Ecological Infrastructure

城市分层　City Layers

通过生活污水处理系统改变地表景观，增加邻里绿地及居住区绿色开放空间，与现存公共绿地结合，丰富城市绿地层次，创造市民游憩交流空间并改善城市面貌及微气候。

中心城区半径约2km，应提倡慢行交通。通过建立自行车道，连接居住区绿地、公园、农田、遗产点以及城市周边区域。在城市内部，减小部分机动车道宽度，两侧增设自行车道，利用现状步道或增设自行车道串联居住区与公园湿地池塘；在城郊，利用现存绿道并新建穿越农田的自行车道。通过建立自行车道系统，使市民在工作、游憩路途中体验多样的自然人文景观，并在观察生活污水净化回用过程的同时提高环保意识。

Ponds contribute to open space and ameliorate microclimates while treating grey water. Improved sewage treatment facilities made it possible to develop more public recreation and communication spaces.

Combined with existing public green space, this water treatment infrastructure helps to improve the urban landscape and micro climate.

The radius of the city centre is about 2 km for which chronic traffic should be advocated. Through the establishment of bicycle lanes, residential green space, park, farmland, heritage point and supply areas would be connected. In inner city, part of the traffic lane would be narrowed and bikeway would be added on both sides. Residential area and wetland park ponds would be connected by existing walkways or new bikeways; In suburb, new bikeways would be set through the farmlands. Through the establishment of bikeway system, citizens would enjoy and learn from diverse natural and cultural landscape on the journey to work and recreation.

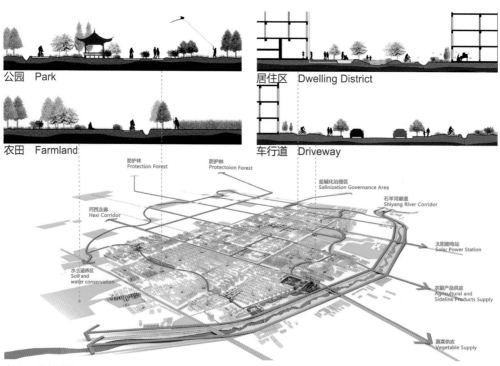

公园 Park

农田 Farmland

居住区 Dwelling District

车行道 Driveway

中观尺度鸟瞰图　Mesoscopic Bird View

微观设计
Microcosmic Design

微观设计选取武威市区一块包括多种景观特征区域，如河道、农田、居住区、绿道的场地。设计将原有的滨河机动车道移至建筑后方，使整个区域内行人可安全自由活动，同时打造一条穿梭于开敞、闭合空间的绿道。在绿道两侧布置水渠，用净化过的生活污水来浇灌乡土植物。另外，将原有的河道改成锯齿状，可以减少泥沙对河岸的冲刷，防止水土流失。场地北侧原为废弃地，经设计后改造为一片都市农业观光地，同时兼具蔬菜供应、文化教育的功能，并运用适合当地环境的浇灌方式作为农业示范区。

The site was selected in Wuwei which includes a variety of landscape character areas such as river, farmland, residential areas and greenway. The original driveway was relocated behind the buildings to make sure that citizens enjoy a safe space and free of vehicular traffic. A green way crossing the open and closed spaces was also established. Treated sewage flows through channels built on both sides of the green way is used to "fertigate" (irrigate with nutrient-rich water) native plants. In addition, the shape of the river bank was changed into a saw-tooth ladder shape to prevent soil erosion. The northern area was abandoned land before it was designated for urban agriculture sightseeing. It also functions as vegetable supply area, educational space, and agriculture demonstration area where the suitable irrigation methods for the local environment were adopted and displayed.

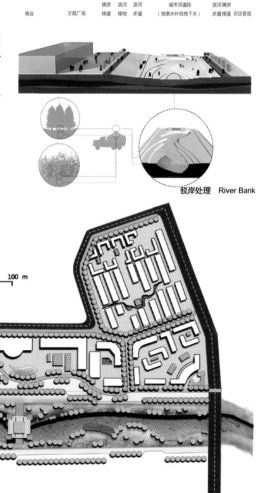

驳岸处理　River Bank

微观总平面图　Microcosmic Master Plan

由于石羊河上游的农田开垦和祁连山植被破坏，石羊河流域水土流失严重。河水中带有的大量泥沙导致下游河道的淤积和红星山水库水位抬高。在武威市区内河道的改造中，设计将原有的笔直河道变窄，河岸变为锯齿阶梯状。在增加亲水空间的同时，不影响河道的防洪能力。

弯曲河床的水流在惯性离心力作用下趋向凹岸，使其水位抬高，从而产生横比降和横向力，使泥沙在弯曲河道的横向环流作用下堆积在河岸边。由于地下水高度矿质化和不科学的农业灌溉，武威市北部地区出现严重的土壤盐碱化。沉积的河泥质地疏松、有机质含量高，可用于更换武威市周边的防护林和农田土壤，改善土壤盐碱化问题。从防护林和农田中更换出的土壤可用于河道修葺等工程用途。

Because of the vegetation destruction at upstream farmlands and Qilian mountain, serious soil erosion at Shiyang river basin occurs. A large amount of sediment in the river causes the silting-up of downstream riverbed and Hongya mountain reservoir bed. In the restoration of river in Wuwei city, the original straight river would be narrowed, and the bank of the river would be changed into sawtooth ladder shape. And the increase of the hydrophilic space will not affect the river flood control ability.

Water in bend river flows under the action of inertial centrifugal force to concave bank. Water near the concave bank rises and produces traverse gradient and horizontal force. So sediment accumulates on the banks of the river . Due to the highly mineral groundwater and unscientific agricultural irrigation, serious soil salinization appears at northern area of Wuwei city . Depositional mud is soft and rich in organic matter content which can be replaced the soil of protection forest and farmland surrounding Wuwei city, in order to relieve soil salinity problems. The soil replaced from protection forest and farmland can be used for engineering purposes such as repairing the bank of river.

社区建筑屋顶上放置了太阳能光电板,能有效利用当地的太阳能,并将其转换为电能。居住区设计沿用具有当地特色的"回"字形空间。这种空间对武威当地的恶劣气候有一定的防御功能,并为居民提供了公共活动空间。居住区内设计有生活污水回用系统。生活污水通过管道收集,经过沉淀池后,进入土壤过滤池中过滤。过滤后的生活污水流入自行车道旁的水渠中,用于灌溉城市绿道。灌溉后多余的水将排放到公园现存的水池中,用于观赏和补充地下水。土壤过滤池中种植香蒲、水麦冬等水生植物,池旁种植刺槐、红柳等乡土遮阴树木,减少过滤池的水蒸发。过滤池在居住区内为居民提供了休憩空间。

在场地内部布置小片区节水农田,形成城市生态基础设施中的点状元素,并提供旅游观光、科普教育服务。田中用大棚种植节水作物。农田灌溉用水来自社区的中水。中水收集在净化池中,经过沉淀过滤,由管道引至农田。

Residential area design continued to use space with local characteristics. These spaces could defence bad weather and provide the public space for residents. Domestic wastewater recycle system was designed in the residential area. Sewage would be collected by the pipe, purged through sedimentation tank, then flowed into the soil filter basin. Treated sewage then flowed into the canal along the bikeway to irrigate the plants. The redundent water would then be discharged into existing pools in parks for landscape sightseeing and groundwater recharge. Aquatic plants such as *Cattail*, *Triglochin palustre* would be planted in soil filter basin. Native plants such as Robinia Pseucdoacacia, *Tamarix ramosissima* would be planted beside the soil filter basin to reduce water evaporation. Filter basins in residential areas provide recreation spaces for residents.

There are some water-saved field plots acting as the dot elements in the urban ecological infrastructure in the site. They also function as places for tourism and science education. Many sheds are designed for water-saved crops. Gray water from nearby communities is used for irrigation. The grey water is collected in purification pool, and flows to the filed through pipes, after sedimentation and filtration.

沉淀池　过滤池　沉淀池
Sedimentation Chamber　Filter Bed　Sedimentation Chamber

邻里空间　Neighborhood Space

都市农业　Urban Agriculture

城郊低洼滞水区的改造与利用
Transformation and Utilization of Suburban Informal Low-lying Water Storage Areas

武威市郊湿地公园规划设计——城郊低洼滞水区的改造与利用
Design of Wetland Park in Suburb of Wuwei——Transformation and Utilization of Suburban Informal Low-lying Water Storage Areas

程红济、刘芹芹、赵茜、寇淼、张玥一
Cheng Hongji、Liu Qinqin、Zhao Qian、Kou Miao、Zhang Yueyi

设计项目位于未来城市北部边界，占地720hm²。目前存在诸多问题，如水资源短缺，绿色基础设施匮乏，城市与农村过渡生硬，农村生产模式与生活模式落后。本设计站在城市发展的角度对该区进行综合规划设计，旨在将该区打造成综合生态基础设施与新型现代农业模式的城乡融合性湿地公园。

The planning object is a wetland park located in the north of Wuwei suburban area covering over seven hundred and twenty hectares. We find that there are several problems in this area, such as the lack of green infrastructure, the severe shortage of water resources, the rigid transition between urban and rural areas, and the undeveloped production mode and living mode of the rural area. The design is based on the development of the city and intended to develop this area to be a wetland district that combine urban and rural areas which include ecological infrastructures and modern agricultural areas.

经过实地调研发现，设计场地存在诸多问题：水资源严重短缺；绿色基础设施匮乏；城市与农村过渡生硬，彼此间的联系不大；农村的生产模式与生活模式还处于十分传统与落后的阶段。基于项目所处的特殊地理位置（未来城市的北部边界地带）以及对该地区的地形地貌、自然资源、气候特征、人文风土的深入了解，本方案抛弃了传统的湿地公园设计思路，站在整个城市发展的角度对该地区进行综合规划设计，旨在将这片区域打造为综合生态基础设施与新型现代农村模式的城乡融合型湿地公园，并提出了城市、湿地、农业彼此融合的模式，不仅在景观上形成一个有机的整体，在功能上还能彼此服务，其中设计重点有五部分：（1）根据场地的不同情况，将该地区划分为六大景观特征区域：西部新型绿色城市区域，南部城乡融合区域，东南节水型城市湿地公园区域，中部体验式农业休闲区域，上中部农业景观湿地区域及北部新型现代农村区域。（2）以公交车、自行车为主的道路交通体系：主干道与城市内部相连接；湿地区域主干道以绿色公交系统为主导；四通八达的自行

鸟瞰图 Birdview

车道将六大景观特征区域串联起来，并深入到城市内部。（3）根据人的出行距离与实际需求，在场地中设计不同尺度与功能的城市开放空间，如小公园或广场等，并与场地中其他部分的绿色系统串联为一个大的体系。（4）湿地形式为潜流式与表流式，湿地景观用水全部来自湿地南侧污水处理厂，尽可能地保护了石羊河的水源，经过湿地净化作用的污水形成了多样的景观。（5）发展新型现代农村模式，形成了粮食、水、资源自循环的自给自足的农村模式。节约了水资源，推进了农村的发展，延续并创造了更为激发人交往的新型农村社区空间。

The planning object is a wetland park located in the north of Wuwei suburban area covering over seven hundred and twenty hectares.Through a vast investigation and research, we find that there are several problems in this area, such as the severe shortage of water resources, the rigid transition between urban and rural areas, there is little connection between the two, and the production mode and living mode of the rural area being very traditional and undeveloped. The design is intended to develop this area to be a wetland district that combines urban and rural areas including ecological infrastructure and modern agriculture. The keynote of this design includes five points. 1. The area is divided into six landscape character areas according to the different situation of the site. 2. Buses and bicycles are prior in the transportation system. 3.The design creates different scales of open space according to the traveling distances and the need of people such as small parks and square. 4.The wetland can be classified into two categories ground stream wetland and underground stream wetland which saves a lot of water. 5. A modern agricultural mode is developed which is intended to create a self-sufficient rural community in which the food, water, and other natural resources are circulated.

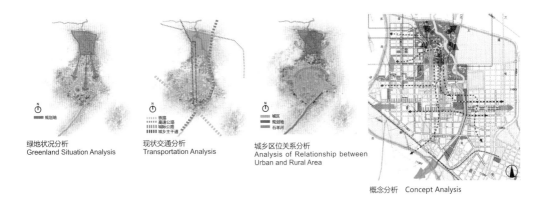

绿地状况分析 Greenland Situation Analysis

现状交通分析 Transportation Analysis

城乡区位关系分析 Analysis of Relationship between Urban and Rural Area

概念分析 Concept Analysis

现状城乡关系剖面示意图 The Situation Map of Relationship between Urban and Rural Area

规划城乡关系剖面示意图 The Future Map of Relationship between Urban and Rural Area

宏观分析
Macroscopic Analysis

在宏观层面，通过对规划地与武威市区的分析，设计者发现目前在宏观尺度上规划设计主要解决以下几个问题：（1）城市与规划地之间彼此不容的状况。（2）现状城市绿地紧缺，人均绿地率低且可达性相对较差的问题。（3）城市与规划地之间通达性差的问题。据此，设计者提出了将规划地与城市现状绿地连接起来，形成一个融于城市的绿地系统战略。

A macro-scale of Wuwei and the site revealed several design problems. 1. the physical separation between the city and the planned area. 2. the lack of accessible open green space relative to the population. 3. limited articulation of the city and the designed area.

通过对设计场地人群类型及主要行为的分析，发现农民的交往模式主要有以下几种（详见下图）。这些行为模式体现了当地人的一种亲密的交流模式，也是进行场地设计时重点要考虑延续的内容。通过对场地现状分析，提取主要的设计要素：太阳能、风能、建筑形式、农田、河流等。

在进行规划设计时，从场地现状中获取灵感，并综合考虑武威市现状以及未来可能的发展潜力，旨在为当地人提供一种适合当地发展的规划模式。

According to the analysis of the people's types and behavior in this area, it is found that the activity modes of farmers contain several

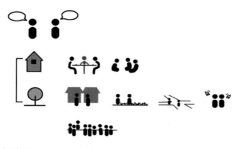

概念图　Concept Map

场地现状照片
The Photos of the Designing Area

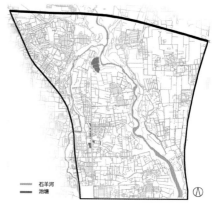

河流现状图
The Situation Map of the Rivers

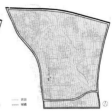

道路现状图
The Situation Map of the Roads

建筑类型现状图
The Map of the Buildings Types

建筑布局现状图
The Map of the Building Locations

LCA 现状图
The Situation Map of the LCA

1. 新型社区
2. 农业湿地景观
3. 农业体验区
4. 城市湿地景观
5. 果园
6. 绿色城市社区

1. New Rural Community
2. Agricultural Wetland Park
3. Experiential Agricultural Area
4. Urban Wetland Park
5. Orchard
6. Green Urban Community

总平面图　The Design Plan

kinds (as the following pictures). These activity modes reflect close relationship among the local people and that is what it is supposed to remain in the design. Through an investigation to the site, the design elements that are proposed are solar energy, wind energy, the patterns of buildings, farmlands and rivers.

In the design process, we were inspired by the site, and considering Wuwei's current conditions and development potential, we aimed to provide the planning mode best suited for local conditions.

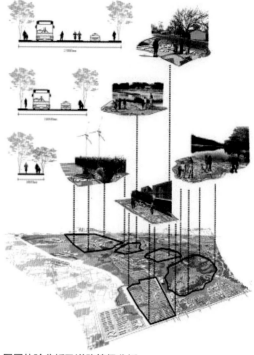

居民体验分析及道路等级分析
Resident Experience and Highway Grade

湿地景观效果图
Wetland Sketch

公园剖面图　Section

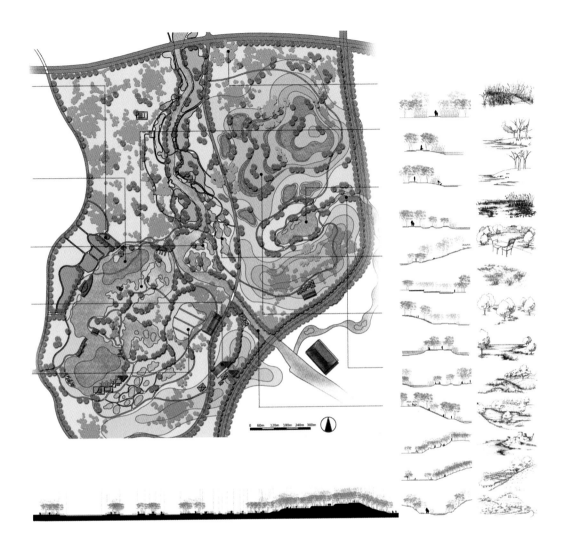

　　湿地公园用水全部来自南侧污水处理厂，最大限度地保护石羊河水源，节约水资源；创造了12个丰富多样的景观特征区域，以适应不同季节及水量变化；分为潜流与表流两种形式，节约水资源同时创造不同景观。

　　The water in the wetland park is all sourced from the sewage treatment plant which saves water and protects Shiyang River. The wetland park is designed with twelve landscape theme zones, The park looks and acts differently depending on the season and water availability. The wetland includes both surface and underground drainage flow which saves water and creates landscape variety.

针对该地区的湿地设计，主要有两个主要概念：（1）山水格局的利用，通过挖湖剩余的土营造高起的地势；（2）利用现有自然景观肌理（滩涂），保留并运用到湿地景观中。

该设计注重人的体验和季节性的景观变化，成为一个随时间变化、生长在场地上的湿地公园。

According to the area of wetland design, there are two main key concepts, 1. landscape use pattern, by digging lake high residual soil to build up the terrain; 2. the use of the existing natural landscape texture (beach) reserved and used in wetland landscape.

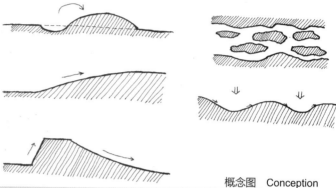

概念图　Conception

透视图　Perspective

新型现代农村模式设计要点：（1）采用现代化的农业生产模式及现代农业技术，形成自给自足的农业社区单元。（2）延续场地肌理及农民生活习惯，创造更为丰富的鼓励交往的农村社会空间。（3）根据对当地现有建筑空间及形态的研究，创造出一种新型的集约化建筑空间，既满足当地气候以及人文的需求，同时精简了土地的利用形式，使当地使用者能享受到生长于该地区又高于该地区现有的建筑空间体验。

The keynote of the modern rural mode design is as follows.

1. The design applies modern agricultural technology to create a self-sufficient rural community.

2. The design keeps the original patterns and living customs of the site, and creates a more sophisticated living space that encourages people's communication and promotes security.

3. Based on the space and form of the existing buildings, a new type of intensive architectural space was created to adapt to the arid local climate and to meet people's needs.

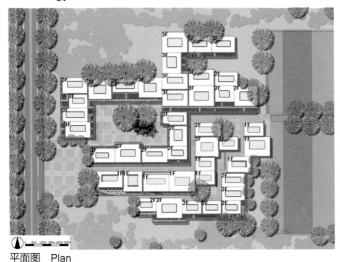

平面图　Plan

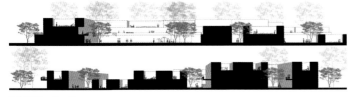

社区系统图　Community System

剖面图　Section

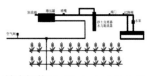

技术细节图　Technical Details

本地区传统的农村街区模式为各个家庭封闭于院落，村落中缺乏较大型的公共活动空间，但其街道朝向利于采光避风。提取传统街区的线性元素，拓宽街道空间创造交流机会。设计新型农村街区的围合空间为村民的交流及娱乐活动创造积极的空间环境。新型农村街区进行集约化设计，北部建筑三层，向南层数依次降低，以营造无风的活动庭院。二三层设计晾晒走廊满足农民生活需求。新型农村街区的空间尺度丰富，通过不同尺度的空间串联，同时通过高差创造村民交流的机会。

In traditional rural communities, every family has their own enclosed courtyard which reduces communication among neighboring households. But the orientation and the forms of the traditional blocks are good for daylighting and avoiding north wind. Picking out the linear element from the traditional rural community, and broadening street space can create more communication opportunities. More closed space is designed in the new modern rural community, so villager can get better opportunities for entertainment and communication. The new rural community is intensive. The northern buildings are mostly three floors, and the buildings are lower to the south, thus creates a pleasant courtyard. Wide balconies are designed in the second floor for drying grins, satisfying farmers need at the same time.

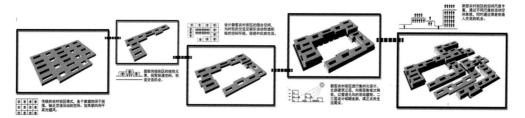

概念生成　Design Concept

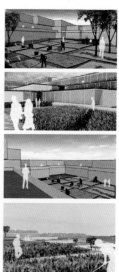

透视图　Perspective

城市公园改造提升
Upgrading of the Existing City Park

N次方公园——城市公园改造提升
N-th Power Park——Upgrading of the Existing City Park

向林森、李绪文、曹安康、马静薇
Xiang Linsen, Li Xuwen, Cao Ankang, Ma Jingwei

当前城市土地都被赋予特定的功能，利用方式单一，效率低下。本方案通过在西郊公园这块土地上叠加多种功能，使这块城市土地利用复合而高效。希望以此唤起人们对城市单一土地利用方式的关注和思考。

Each piece of urban land is endowed with a specific function, resulting in low service efficiency. This project is aiming at mixed land use by proposing a variety of functions in the design of western suburb park. We hope to arouse people's attention and thought to the current low efficiency urban land use pattern.

1．背景
本组的研究选题为武威市西郊公园的更新设计，西郊公园始建于1980年，面积18hm^2，现在的西郊公园使用人数多，服务范围广，但由于建造年代久远，基础设施建设不完善，公园迫切需要一定程度的改造以满足市民新的需求。

2．设计理念
在城市化建设的过程中，原本具有多种功能的土地利用方式日益被单一的城市土地利用方式所取代。一张又一张的土地利用规划图将原本高效的土地利用变得低效而且乏味。在城市环境恶化，土地日趋紧张的今天，城市中的每一寸土地都应该发挥它的多重属性，而不是被规划人员简单地定义为某一种功能的用地。

本方案将西郊公园的设计概念定义为"N次方"公园，即这个公园不仅仅是一块绿地，而是一块具有生产、生活、生态、游憩、文化、教育、能源转换等多重属性的公园，这些属性的叠加又使它具有更多的可能性，就如同有N个传统意义上的公园同时叠加在这块土地之上。希望以此来唤起人们对土地及土地利用方式的关注。

3．设计途径
通过调研发现公园最具特色的景观元素是公园北部的菜市场。菜市场的存在已经在一定程度上改变了传统意义上公园的使用方式，并且影响到了公园的空间划分。因此，本方案将菜市场作为设计的切入点。

以上述切入点为基础，对整个公园的结构和功能进行了二次梳理，将游憩、休闲等传统意义上的公园功能同生态、能源转换、粮食生产、商业等功能进行融合，环环相扣，层层叠加，使新的西郊公园以多重身份参与到城市的运作中去。

1. Background

This project is about the renewal design of Western Suburb Park, which locates in Wuwei, Gansu province. Founded in 1980, the park covers an area of 18 hectares, serving a large area. However, as this park was constructed decades ago, the infrastructure lags behind. It is an urgent demand to transform the park to keep up with the rapid urbanization in Wuwei.

2. Design Concept

In the process of urbanization, the original mixed land use is replaced by a single urban land use pattern. The farmland used to be productive, living and ecological. But now, functions are assigned to separate land. The land use plans gradually become inefficient and boring. Facing the increasingly serious crisis of resources and environment, every piece of the land should make the most of its multiple attributes and maximize the value.

Park is supposed to be not only a piece of green space, but also a place for production, living, ecology, leisure, culture, education, energy conversion. Therefore, we propose the concept of "N-th power" park, which means more possibilities and multiple functions, indicating a mixed land use pattern.

3. Design Approach

In the field trip, the food market which lies in the northwest corner of park turns out to be the most exciting feature. As the market has affected how the space being divided and used, we decided to start the design with the market. The park will play a very important role in the city development by combining the functions such as recreation, ecology, energy conversion, food production and commercial.

公园位置　　　城市发展趋势分析
Site location　　City Develop Trend Analysis

农田、公园和菜市场服务范围分析
Farmland, Park and Market Service Area Analysis

宏观分析
Macroscopic Analysis

武威是中国历史文化名城，全国优秀旅游城市，古丝绸之路的要冲，具有较高的文化旅游价值。武威市夏季短而炎热，冬季长而严寒，春秋季表现不明显，年、日温差较大，降水少，蒸发量大。

未来整个武威市主要向西北及东南方向发展，向西扩展的面积很小。

未来的绿地系统主要是新规划的面积较大的湿地公园和海藏公园，但二者的定位都不是服务市中心的综合性公园。

从城市及绿地的发展趋势看，西郊公园将一直处在城市的核心位置，成为沟通南北绿地的关键点。如果将武威市的绿地系统比作一根扁担，那么西郊公园就是这根扁担的中间着力点。

农业用地逐渐被城市扩展所蚕食，城市发展占用了大量的良田。

As one of the Chinese famous historical and cultural cities and the national excellent tourism cites, Wuwei is the communications hub of the Siik Road, and has a great value of its cultural tourism. The summer there is short and hot, the winter is long and cold. The spring and fall are not obvious; Annual and diurnal difference in temperature is large, precipitation is low, and evaporation is great.

In the future, Wuwei city will mainly expand to its northwest and southeast, the area of the west extension is very small.

A new wetland park and new Haizang park will be built according to the green space system plan of Wuwei. Both of them are not comprehensive park aiming at center-city service.

From the city and the development trend of green space, the western suburbs park which would continue to be the heart position of the city will become the key point between north and south green space. Take Wuwei's green space system as a lever, the western suburbs park would be the fulcrum of it.

Agriculture land is replaced by the rapidly expanding urban area. As a result, city's development has sacrificed many good farmland. Obviously, it is unsustainable.

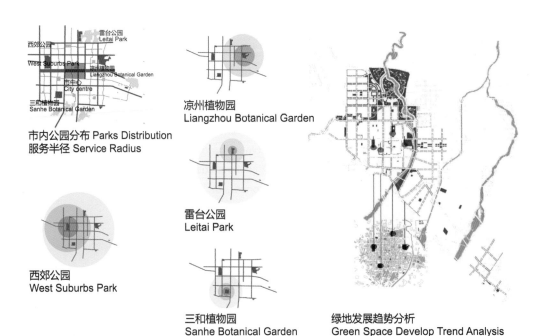

市内公园分布 Parks Distribution
服务半径 Service Radius

西郊公园 West Suburbs Park

凉州植物园 Liangzhou Botanical Garden

雷台公园 Leitai Park

三和植物园 Sanhe Botanical Garden

绿地发展趋势分析 Green Space Develop Trend Analysis

中观设计
Mesoscopic Design

现有的空间形态将各个活动分隔开，但是人的活动使得这些阻隔正在被打破，而人们渴望的正是这样的空间：各个活动之间可以相互交流，从而激发更多的创造力和活动。现场观察到，买菜这一日常活动形成了人们和公园的关系纽带，人们提着菜篮子穿过公园，或带着孩子去早市遛弯，或在公园里面悠闲地散步，看来来往往的人们穿行。菜市场成为公园活力的源泉之一，也是引发多种可能性活动的动力源泉。

The existing configuration of space separates people's activities by function such as shopping or recreation, however, it is people's activities that break these separations creating integrative spaces. People want to communicate across the spaces that divide them in order to arouse more creativity and activity. Based on our observations, many people walked through this park to go to the market with their families. Others would prefer to just walk and rest in the park. Grocery shopping, one of the daily activities taking place in the market, is becoming the connection between people and the park.

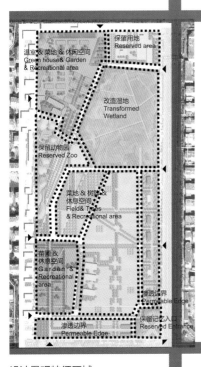

现状景观特征分区
Current Landscape Character Area
注：灰绿色表示景观特征不明显区域
Note: Green grey area means easily ignored areas

公园里的活动
Activities in the Park

设计景观特征区域
Design Landscape Character Area

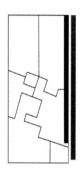 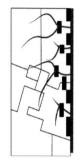

公园东面的建筑将公园和城市完全分开，公园成了人们生活的偶然性因素。我们的设计意向是将公园的边界打破，让外面的生活进入公园，让公园融入生活。

On the east side of the park, buildings separate the park and the city completely making the park a contingency factor of people's life. We want to open-up the park boundary to let the life outside become an integrated part of the park landscape and infuse the park with life.

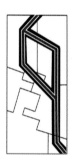 

公园现在更像一个通往菜市场的快速通道，在这条通道上没有休闲、娱乐和交流。在设计过程中，我们希望打破这种单一的公园使用方式，让人们进入，使用，并享受公园！

Park is more like a fast road to the vegetable market now without leisure, entertainment and communication. We hope to break the dull way of the park use, let people enter, use, and enjoy the park!

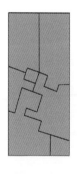 

现在公园的植被被均匀地种植在场地中，没有节奏，没有韵律，没有方向，也没有惊喜。我们希望我们的设计能够最大限度地丰富人们的景观体验。

The trees in the park are now uniformly planted in rows; there is no rhythm, no direction, and no surprise. We want our design to enrich the landscape experience of visitors.

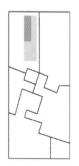 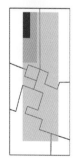

公园的北部配置有一处菜市场，本设计力图给人们创造一种全新的、舒适的、卫生的、充满乐趣的蔬菜市场新体验。

There is a vegetable market in the north of the park now. Our design intent is breaking the traditional concept of the vegetable market, giving people a new, comfortable, healthy, and interesting experience.

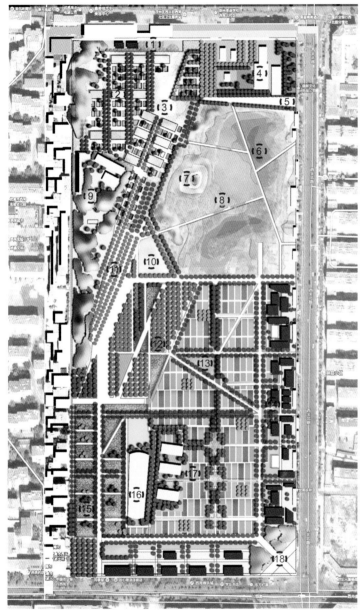

图例 Legend

1 树阵过渡区域 Trees Transition
2 温室、田地混合区
  Mixed Area of Green House and Field
3 活动广场 Square
4 少年宫 Reserved Youth Plaza
5 入口 Entrance
6 湿地 Wetland
7 山丘 Hills
8 休息平台
  Recreational Platform
9 动物园 Zoo
10 草坪 Lawn
11 保留树林 Reserved Trees
12 保留树林与苗圃
   Reserved Trees and Nursery
13 儿童游戏场所 Playground
14 葡萄架 Pergola
15 苗圃 Nursery
16 保留溜冰场
   Reserved Skating Rink
17 菜园 Vegetable Garden
18 记忆入口 Memory Entrance

保留建筑 Reserved Buildings
新增建筑 Additional Buildings
保留植被 Reserved Vegetation
苗圃 Nursery
菜地 Vegetable Garden
------> 自行车道 Bikeway

总平面图
Master Plan   0 20 40 80 m

| 55

微观设计
Microcosmic Design

水系统设计积极回应武威市干旱缺水的现状,将公园周边社区生活污水收集到公园,在动物园下安排污水处理温室,经预处理后进入湿地生态系统,经过潜流湿地、表面流湿地层层净化进入中心水体,水处理系统本身就是优质的景观资源,与不同的景观单元结合后给人以不同的景观体验,污水净化后被用作观赏、灌溉、冲洗厕所等。在都市农业区,应用生物净化技术净化污水,成为公园另一道独特景观。

Water system design responds to the status of drought of Wuwei positively. To restore the natural water cycle, the waste of the surrounding communities is collected to the zoo, after pretreatment by underground sewage treatment greenhouse, the water runs into the ecological wetland for further purification. Water treatment system itself is excellent landscape resources, after combined with different landscape units, it shows a different landscape experience. The treated water is used for landscape, irrigation, and flushing toilets. In the area of urban agriculture, living machine water purification technology would be applied which will become another unique landscape.

1. 记忆入口
公园的入口现状非常具有年代感,植被丰

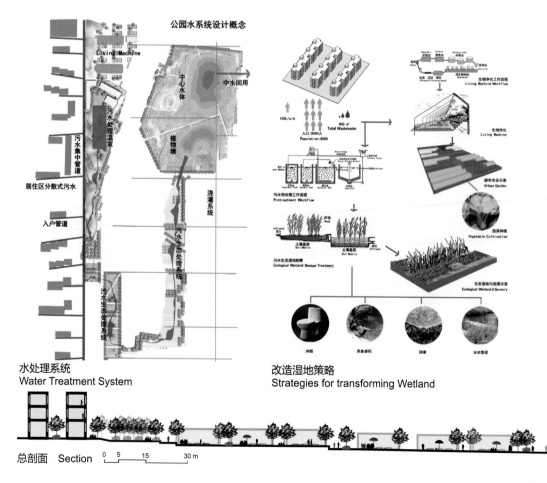

水处理系统
Water Treatment System

改造湿地策略
Strategies for transforming Wetland

总剖面　Section　0　5　15　30 m

富，因此本地块的设计主要是希望保留这种年代感并将其强化。空间尺度及大部分植物都被保留。同时，新增了一些具有年代代表性的元素，例如露天电影、跳格游戏、滚铁环等场地，并利用这些元素布置组织起多层次的空间，使人获得由浅入深，由淡到浓，真实丰富的景观感受，唤起每个人心底的回忆。

1. Memory Entrance

The current entrance invokes a strong sense of tradition and the vegetation is lush, the design wants to maintain and strengthen this feeling. Based on the present space scale, most of the plants are reserved, traditional recreational elements have been added such as open-air movies, jumping games, and hoop rolling. These elements are decorated to organize multi-level space and arouse traditional and childhood memories while people enjoy physical scenery.

2. 菜市场、温室大棚和苗圃

在城市能保有乡村景观元素难能可贵。在公园的西北角开辟菜园，建立温室大棚，开放原有的菜市场与原本封闭存在于公园的苗圃，并与茶室相结合，极具当地特色。都市农业景观成为公园的主导，而市场也溶解在了这一景观基底之中。这种城乡融合的模式将让人耳目一新。

2. Market, Greenhouse and Nursery

Maintaining rural landscape elements in city is commendable. Vegetables are cultivated in northwestern area, greenhouses are established. The local market and previously isolated children's nurseries are opened up and combined with a teahouse to add local character. The urban garden landscape dominates the whole park, the market is dissolved in the landscape.

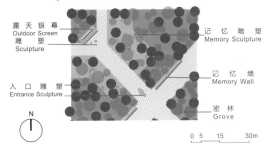

记忆入口　Memory Entrance

记忆墙　Memory Wall

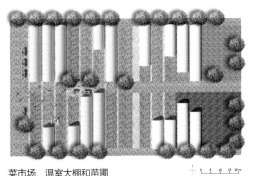

菜市场、温室大棚和苗圃
Market, Greenhouse and Nursery

温室大棚和苗圃
Greenhouse and Nursery Perspective

历史街区保护与更新
Protection and Renewal of the Historic District

土城——历史街区保护与更新
Soil City——Protection and Renewal of the Historic District

袁远、倪冰、刘玥、谢芳丽、蒋晓峰
Yuan Yuan, Ni Bing, Liu Yue, Xie Fangli, Jiang Xiaofeng

武威作为西北地区重要的历史名城，历史城区如何保护和发展也是亟待解决的问题。本方案以"土城"为主题，通过"下沉城市"设计、生态古城设计以及综合使用用地等方式，探讨合理更新和健康发展的新模式。

近年来，武威的城市发展过程中，历史古迹毁坏严重，大量具有民族特色的住宅、古遗迹被夷为平地，修建了商铺、住宅楼，对历史名城保护产生了极大的破坏。本规划在此背景下，对武威市区范围内的罗什寺、古钟楼、文庙等三个片区进行研究，以三大片区为核心，通过旧城改造、文化创新、现代商业引导等方式，增加历史街区活力，辐射周边地区，延续当地历史文化。并以"土城"为主题，可持续发展为原则，对古钟楼片区进行了场地设计。场地主要分为两大部分："旧土城"和"新土城"。旧土城以建筑修葺和立面改造为主，新土城整体下沉5米，形成一个"下沉城市"。

Wuwei is an important historic city in northwest China. How to protect and develop historical district in Wuwei is also a problem to be solved in a rush. This plan takes "soil city" as the theme, and explores a new pattern of rational renewal and healthy development, by the idea of "sinking city", design of ecological city and application of mixed land use.

In recent years, with the development process of Wuwei, historical sites have been destroyed seriously. A large number of national characteristic ancient residence and monuments were razed to the ground to build the shops, residential building, which has a great damage on the historic city protection. In this context, the Ancient Bell Tower area, the Luoshi Temple area and the Confucius Temple area have been studied. Taking these three areas as the core, this plan tries to improve the vitality of the historical block through the renewal of the old city, cultural innovation and guidance of modern business, in order to stimulate the surrounding area and facilitate the continuation of local history and culture. With "Soil City" as the theme and sustainable development as the principle, the site design of Ancient Bell Tower area will be reinvigorated. The site is divided into two parts: the "old soil city" and the "new soil city". The former has for the repair and facade renovation of buildings while the latter is 5 meters lower, forming "sinking city."

宏观分析
Macro analysis

武威是丝绸之路上的一个重要城市，自古以来受多种文化的影响，历史文化悠久并具多样性。武威市的土壤类型主要是栗钙土和黑钙土，适宜种植当地的植物以及作为建造当地房屋的材料。古钟楼片区作为武威市中心的一个重要文化点，通过合理的规划设计，逐渐扩大其文化影响力。

Wuwei is an important city on the Silk Road, which has been affected by a variety of cultures since ancient times. So it has a long history of diverse cultures. Chestnut soil and chernozem soil are the main soil type, which are suitable for local plant as well as the local building materials. Ancient Bell Tower area, as an important cultural point in the center of Wuwei, can gradually expand its influence in the culture through reasonable planning and design.

场地分析
Site Analysis

对古钟楼片区进行了使用者活动行为的分析、温度与风的分析以及景观元素的分析。基于这些分析，提出了"土城"的设计概念，并对后期的设计起着一定的指导作用。

Based on an analysis of people's activities, temperature and wind patterns along with and landscape elements in the Ancient Bell Tower area, we have proposed the "Soil City" as the design concept which guides the later design.

区位分析
The Location Analysis

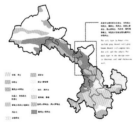

土壤类型分析
The Soil Type Analysis

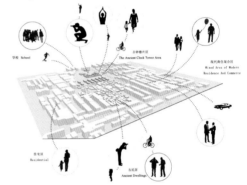

活动行为分布图　The Behavior Pattern

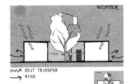

冬季热量分析
Winter Heat Analysis

夏季热量分析
Summer Heat Analysis

文化分析
The Cultural Analysis

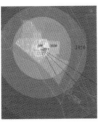

影响力预测
The Influence Prediction

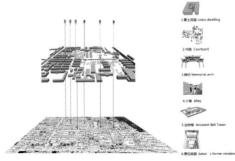

景观元素分析　The Landscape Elements Analysis

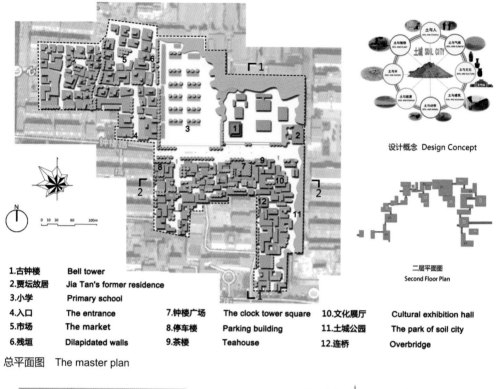

1.古钟楼	Bell tower			
2.贾坛故居	Jia Tan's former residence			
3.小学	Primary school			
4.入口	The entrance	7.钟楼广场	The clock tower square	10.文化展厅	Cultural exhibition hall
5.市场	The market	8.停车楼	Parking building	11.土城公园	The park of soil city
6.残垣	Dilapidated walls	9.茶楼	Teahouse	12.连桥	Overbridge

总平面图　The master plan

1-1 剖面图　1-1 Section

2-2 剖面图　2-2 Section

中观设计
Mesoscopic Design

旧土城：位于场地西北方向，原为黄土民居集中区。大部分建筑保存较为完好，设计以建筑修葺和立面改造为主，提高居住质量；增加绿地和开敞空间，丰富居民生活；完善道路交通，增强与周边地区的联系。

新土城：位于场地东南方向，现为已拆除建筑的空地。设计将这片区域整体下沉5m，形成下沉空间内的建筑的一层屋顶直接与周边地面层相连，人们可以直接走上"二层"步行空间。通过下沉式设计，使得该区域层次多变，极大地丰富了空间；降低了该区域建筑体量，突出古钟楼的宏伟；形成区域微气候，使得该片区能够冬暖夏凉；挖掘出来的土方可用于新土城的建筑材料；"下沉城市"如同扎根于土地，从土里生长出来一样。整个片区一层和二层都是以步行和自行车为主，为混合居住区。

Old Soil City: northwest of the site, is a residential site with buildings constructed of loess which are well preserved. The design is given priority to building repair and facade renovation, to improve the quality of living, increase the green space and open space, enrich people's life, optimize road traffic and strengthen links with the surrounding areas.

New Soil City: southeast of the site, as dismantled building space at present. The region is sunk 5 meters below grade. The roof of single-story buildings and surrounding areas are interlinked in the sunk space so that people can go directly to the walking space at the second floor. By sunk design, the region's levels are changeful, which can enrich space, highlighting the Ancient Bell Tower through reducing buildings' size,and form micro climate that is warm in the winter and cool in the summer. Excavated soil can be used as a construction material. The "Sunk City" grows out of the soil and the whole site encourages walking and cycling through mixed-use residential areas.

微观设计
Microcosmic Design

墙根公园
The Wall-foot Park

墙不是单纯的阻挡物，墙边区域可以成为很好的交流场所。墙根公园除一般公园所具有的休憩、亲近自然功能外，还增加了沿墙步行过程中交流的可能。

Many stories happened along the wall in China.So the wall-root park is not only a place for resting and closed to nature,but also for communication during walking alongside the wall.

墙根公园鸟瞰图
Bird's View Perspective of the Wall-Root Park

长桥   The Long Bridge

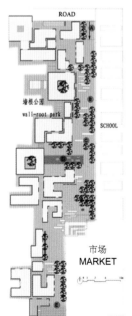

铺装细节
The Details of Pavement

墙根公园平面图
The entrance into the Wall-foot Park

公园入口   The Entrance into the Wall-foot Park

主街
The Main Street

主街的设计思想是把广场串联成路，即每个建筑前都会形成一定的休息空间，可以给使用者更多选择。钟楼对面的广场尺度最大以突出古钟楼的高大雄伟，其他小广场根据高差和用途分别设计，具有不同的形态，极大地丰富了街道空间。街道的设计还充分考虑了遮阴效果、节能要求和历史文化的表现，通过树池座椅、风能、太阳能路灯、自由摊位和泥塑的应用使土城的街道更具特色。

The design idea of the main street is to connect the existing chain of small plazas together into a pedestrian street with resting places giving consumers more choices of places to sit. The plaza opposite the Bell Tower is designed reasonably to underscore the significance of the Ancient Bell Tower while other small plazas are designed according to the elevation and usage. A variety of plaza shapes enrich the street space. The design also gives full consideration to the shade effect, energy-saving requirements, and the influence of history and culture. Features such as tree pool seats, wind and solar-powered street lights, vendors and clay sculptures make the streets of the soil city more attractive and interesting.

主街　The Main Street

主街效果图　Bird's View Perspective

主街及其周边夜景效果图
The Night-view Perspective of the Main Street and the Surroundings

二层夜景　The Night-view of the Second Floor

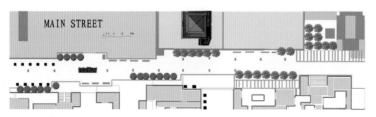
主街平面图　The Master Plan of the Main Street

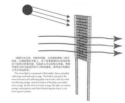
街灯说明
The Instruction of the Street Light

下沉空间
Sunken Space

"新土城"整体下沉5m,形成独特的下沉空间。上下两层都可以步行和骑自行车。下沉地面的街道上是特色店铺、公共建筑、居民住宅等组成的混合空间。并且通过街道布局的设计,丰富了街道空间,形成了社区文化。一层的屋顶也是通行空间,与周边具有连通性。行人可以直接进入建筑的二层,也可以在屋顶平台上休憩、赏景。

"新土城"与周边地区的连接主要有三种方式:连桥、缓坡、阶梯。周边的人可以顺利进入到下沉空间的一层或者二层。对于不可通行的边界,通过草坡地或者绿植进行"软隔离",形成良好的过渡景观。

"New Soil City" is sunk 5 meters below grade, forming a unique space. Upper and lower layers are accessible by walking and biking. The lower level includes a variety of mixed uses such as shops, public buildings, and residential spaces. The enriched street layout forms the social environment to enhance community culture. The roof of single-story buildings is also open space with links to the surroundings. Pedestrians can go directly into the second floor of the buildings, or they can rest, relax, and admire the view on the platforms of single-story buildings.

The design of "New Soil City" takes the connection with the surrounding area into consideration in three main ways: bridges, gentle slopes, and ladders permitting people can move smoothly into the lower level or second floor. The space is clearly defined to form a clear transition landscape through the slope or "soft isolation" of green plants.

下沉空间   Sunken Space

下沉空间边缘效果图
The Boundary of Sunken Space Perspective

下沉空间效果图   Sunken Space Perspective

下沉空间效果图   Sunken Space Perspective

上层与下层的联系
Connection between up and down

空间内部天桥的联系
Connection of Bridge Inside

空间内部楼梯的联系
Connection of Ladder Inside

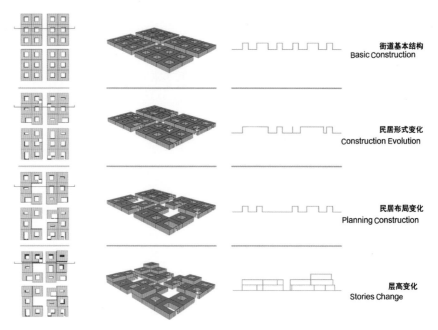

街道结构的演变　Evolution of Street Structure

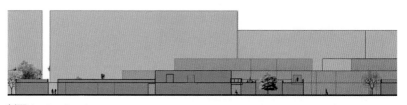

剖面 1　Section 1

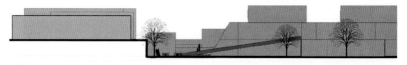

剖面 2　Section 2

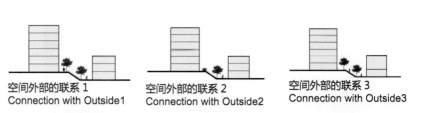

空间外部的联系 1
Connection with Outside1

空间外部的联系 2
Connection with Outside2

空间外部的联系 3
Connection with Outside3

空间外部的联系 4
Connection with Outside4

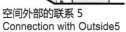

空间外部的联系 5
Connection with Outside5

空间外部的联系 6
Connection with Outside6

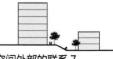

空间外部的联系 7
Connection with Outside7

流动市场——历史街区保护与更新
Moving Market——Protection and Renewal of the Historic District

衡先培、张冰洁、吴梦霞、杨雪、贾琳
Heng Xianpei, Zhang Bingjie, Wu Mengxia, Yang Xue, Jia Lin

从对场地最直接的感知出发提出设计主题，希望在城市历史街区的设计和保护中，不仅保护和延续物质空间的历史性，更能够保护场地上具有当地特色的生活。该主题一方面是对于现在城市的"超级市场"商业模式的一种反思，另一方面也是对于保护街道传统生活和促进人们之间交流的探索。

在宏观层面上，方案希望能够将机动车交通干道尽量移出凉州老城区的范围（约2km²）。在此范围内，用流动市场连接五个主要遗产点，形成步行和自行车主导的街道空间，加强现有的东西大街的商业干道周边街道的商业吸引力，同时通过传统街道空间的塑造为民俗文化、创意产业等商业形式提供机会。而在外围，以现有的滨河景观带为基础建设具有游憩和自行车通行功能的绿道连接整个中心城区。

The theme "Moving Market" derived from the direct perception of site, was put forward in hope that not only the historical material space but also the local lifestyle could be protected and continued in historic blocks. On one hand, "Moving market" was a kind of reflection for "super market" business model in modern city; On the other hand, it was also a kind of exploration for protecting traditional street life and promoting the communication among people.

On the macro level, the project removed motor vehicle traffic from the scope of the old city(about 2 kilometers range centered on downtown). "Moving markets" were connected with five major heritage places by walking and cycling oriented streets to strengthen the attraction of the existing commercial streets which were near to the main East-and-west street, and at same time it can provide various opportunities for all kinds of business, especially folk culture and creative industries. In the peripheral area, greenways for recreation and bicycle were constructed on the basis of the existing riverside landscape.

宏观分析
Macroscopic Analysis

景观结构分析　Landscape Structure Analysis

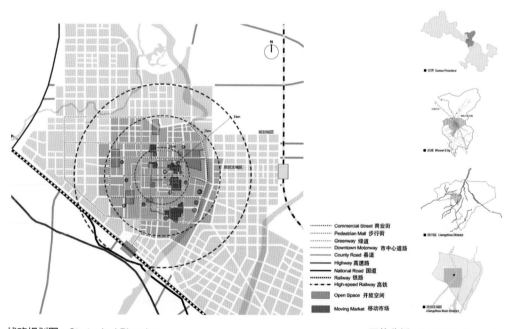

战略规划图　Strategical Planning　　　　　　　　　　　区位分析　Location Analysis

中观设计
Medium Scale Design

设计概念
Design Conception

古钟楼街区的具体城市设计中,建筑肌理源于场地原有的黄土民居,场地东部考虑到与城市界面的衔接安排了办公功能同时对建筑体量进行了放大;场地西部紧靠大云寺、古钟楼和贾坛故居,以保留原有建筑肌理为主;场地南部以居住为主的街区则充分保留街巷空间丰富性。在整个场地的空间布局中,特别注意建筑之间错落形成的街道宽度的变化和小型的街角空间,这些空间和流动市场的商业点结合可以成为邻里、街区和城市中的机会空间。

At the meso scale, the Ancient Bell Tower block continues the building texture of the original loess local-style dwelling houses, but in the east, considering the interface and cohesion with the current city, the volume of buildings with commercial office functions are amplified; Areas close to the Big Cloud Temple and Jia Tan old residence were given priority to the original building texture. Residential blocks in the south have fully reserved lanes contributing to the richness of the space. In the overall layout of space, special attention was paid to the change of the architectural form and street width to create many opportunity spaces, such as small street corner parcels, where periodic market businesses could operate.

总平面图　Master Plan

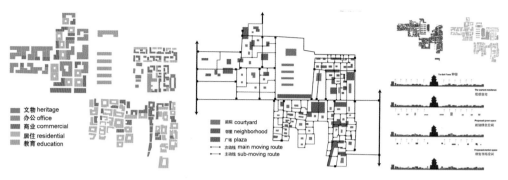

建筑功能分析　Function Analysis of Buildings　　动线分析　Circulation Analysis　　肌理分析　Texture Analysis

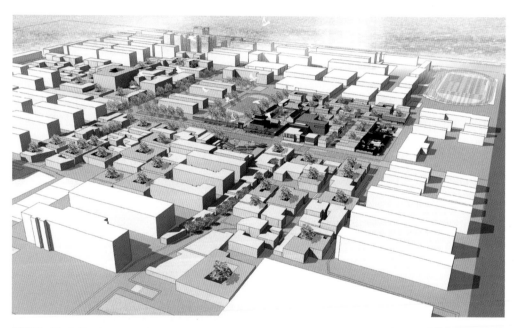

鸟瞰图　Bird's Eye View

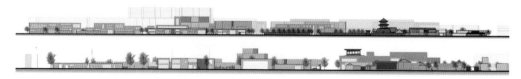

剖面图　Sections

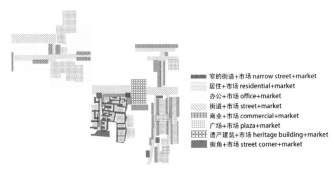

景观特征区域分析　Landscape Character Area Analysis

传统大型商业模式

商业模式分析
Business Model Analysis

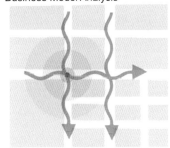

移动市场模式
Moving Market Mode

场地设计
Site Design

　　细部设计主要选取了有代表性的五个节点，探讨了流动市场与不同空间的结合与在此之上可能的活动形式。这五种结合分别是：通过空间划分将大型城市广场和时令性集中商业空间结合；展览建筑和历史建筑间的界面处理和街边空间中流动商业和墙根文化的结合；邻里广场和民俗小吃售卖空间的结合；街角和引入流动市场售卖点的狭窄巷道的结合；居民自发种植的花园和作物自发交换空间的结合。

　　At a micro-scale, five presentative nodes establish the combination of mixed uses: periodic markets, open space and culture. The five combination forms were large city square with seasonal market space concentrations; exhibition buildings with historical features; street space with mass culture; neighborhood squares with small-scale markets; and street corner or narrow lanes with decentralized selling space. More attention will be paid to street life and gardens cultivated with flowers or spontaneous plant transactions.

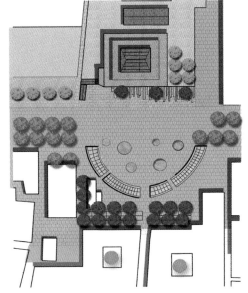

场地1平面图　Plan of Site One

场地一剖面图　Profile of Site One

| 71

场地 2 效果图　Perspective of Site Two

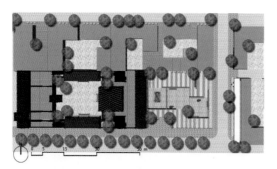

场地 3 平面图　Plan of Site Three

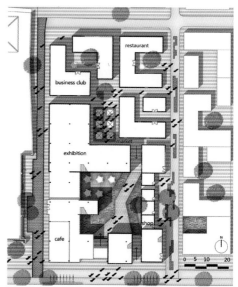

场地 2 平面图　Plan of Site Two

场地 3 效果图　Perspective of Site Three

场地 4 效果图一　Perspective of Site Four, 1

场地 4 效果图二　Perspective of Site Four, 2

场地 4 平面图一　Plan of Site Four, 1

场地 4 平面图二　Plan of Site Four, 2

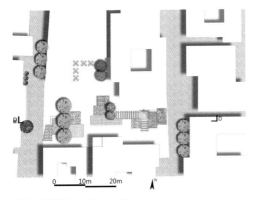

场地 5 平面图　Plan of Site Five

场地 4 剖面图　Profile of Site Four

场地5剖面图　Profile of Site Five

细部设计
Detail Design

场地边界处理示意图
The Solution of Site Boundary

场地5效果图　Perspective of Site Five

废物利用铺装示意图　Recyle Pavement

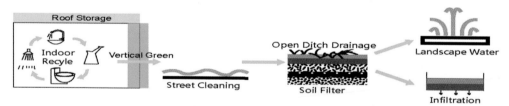

场地水循环概念　Concept of Water Recyle

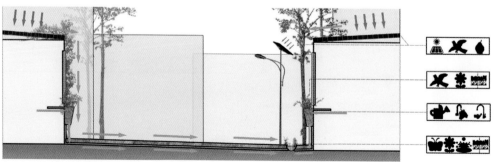

生态与节能方式示意图　Ecology and Energy

城市扩展区可持续城市设计
Sustainable Urban Design of City Expansion Area

城市动脉——城市扩展区可持续城市设计
Urban Arteries——Sustainable Urban Design of City Expansion Area

Andreas Mayor

武威的整个设计愿景是创建一个适合步行和非机动车行驶的城市。在这个城市中，人们进行日常活动的重要节点和场地都可以通过步行或使用自行车到达。设计旨在创造一种运动的相对平等的状态，即使大多数居民买不起车也可以让每个人都在这个城市中自由畅快地活动。

这条河流形成层次丰富的河流景观，为周围居民创造了休憩娱乐的场所。当地耐旱植物形成的旱地景观与季节性的降雨净化了周围环境，塑造了武威市独一无二的美丽风景。该方案利用动车轨道系统连接了城市重要的景观节点，其中主要的停靠点将与新城的广场形成紧密的联系。场地中的广场与交错的街道将会相互融合，通过公共空间的共享来提高使用者的安全感。不同规模，热闹的底层商业店铺提供了形式各异的商业往来，形成了生动活力的街道生活，为当地的人们创造愉悦的城市体验。商业单元、住宅区、独立住宅、市政府、廉租房形成了丰富多样的街道结构。

The overall vision for Wuwei is to create a walkable and bikeable city. In everyday activities, important nodes and sites within the city can be reached by walking or using a bike. The vision aims at creating somewhat of equality in movement, allowing everyone to move around in the city even though most inhabitants can't afford a car.

The river is transformed into a riverscape with different landscape sequences creating a new recreational spine. Xeriscaping clean local grey- and rainwater, at the same time provide an esthetic element to the city, the riverscape give Wuwei a unique city feature. The proposed tram system connects important nodes in the city, the main tram stop is located in close relationship with the new urban square. The square and street crossings in the area are leveled with a shared space principle

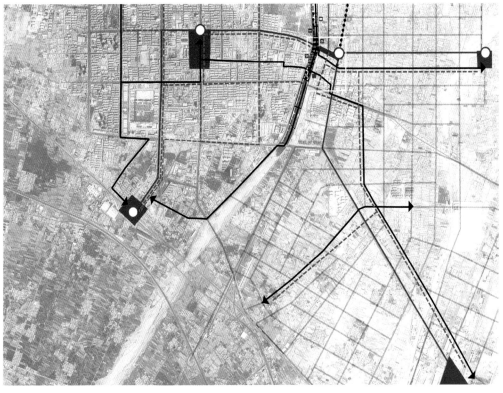

把武威打造成滨河城市
Make Wuwei a River Front City

沿河建立鲜明的城市地标
Create Strong City Identities along the River

城市发展战略：添加新的人行道系统来打破原有超大的街区，创建一个层次分明的街道结构
Urban Development Strategy: break up the superblock, add new pedestrian paths within the blocks and create a strong street hierarchy

重组现有的自行车和行人的街道
Reorganizing the Existing Streets for Bikes and Pedestrians

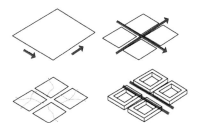

避免市区交通复杂拥堵
Make the Downtown a Traffic Calmed Area

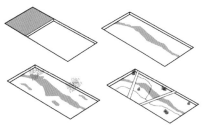

河流策略：开渠动流，河渠收集雨水成为整个城市的一条休闲绿色脊柱
River Strategy: Open up the flow of water through a channel, the river gathers rainwater and works like a recreational green spine for the whole city

to increase the overall safety. Active ground floors with commercial units in various sizes offer a variation of business opportunities while create a pleasant urban experience with a vivid street life. Commercial units, apartments, urban villas and townhouses form different block structures with several housing alternatives for the local.

○ 半径 200m Radius 200m
╱ 路线 Route
◉ 车站 Station

新电车路线
New Tram Route
with Stations

● 城市街区 Urban Block
● 原住街区 Local Block
● 活动带 Activity Strip

区域街区结构
Block Configuration
within the Area

● 商业单元 Commercial Unit
● 公寓 Apartment
● 城市别墅 Urban Villa
● 联排别墅 Townhouse

区域形态结构
Typology Configuration
within the Area

○ 公共空间 Public Space
● 半私密空间 Semi-private Space
● 私密空间 Private Space

公共、半私密和私密空间
Public, Semi-private and
Private Spaces

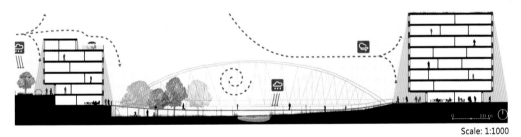

Scale: 1:1000

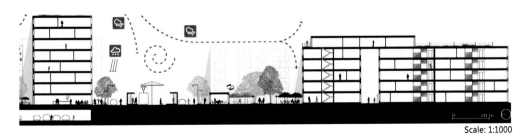

Scale: 1:1000

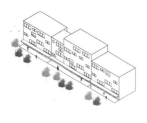  

活动地带 Activity Strip　　城市街区 Urban Block　　当地街区 Local Block

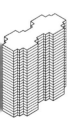

居住塔楼	居住塔楼	板式楼	商业单元	公寓楼	城郊住宅	联排住宅
Residental Tower XX+	Residental Tower X+	Slab	Commercial	Apartment	Urban Villa	Townhouse

城市绿洲——城市扩展区可持续城市设计
Urban Oasis——Sustainable Urban Design of City Expansion Area

Constantin Mihai Milea

项目旨在提供一个武威市区未来水资源精明利用的发展策略。就市区的尺度而言，这个项目所提供的发展策略将使城市的经济得到发展，同时能够为当地人提供地方认同感和充满活力的城市生态系统。本设计的主要概念是"绿洲"，并用这一概念来打造独具特色的社区。对于这片场地的一个挑战是：既要创造一个城市绿洲结构来还原干旱地区的生活，也要提出一个精明用水的具体方案。本设计的另一重点是从街道到建筑形态上创造出多样的，混合利用的空间，这些空间能成为当地人的活动场所。创造一个混合利用的空间能够产生多样化的街道，同时城市绿洲（小尺度的绿地）能够调节小气候，展示随着季节发生变化的景观。邻里尺度上的废水再利用是水系统设计的重点，将废水转化为当地经济和能量，用以重塑绿洲景观。

The projects aim is to provide a strategy design focused on water smart solutions for Wuwei City's future development. Starting from the city's scale the project is working with a vision to provide the framework in which the future site can develop a growing economy while providing local identity and a lively urban ecosystem. The concept of oasis started up as a main idea for the design, shaped to capture the existing neighborhood qualities. The challenge for this site was to create an urban oasis structure to capture the essence of living in an arid area but still to develop a strategy for smart water solutions. Thus the catalyst used to develop the design is focused on creating diversity at the street level extended to the buildings typology creating a mix of space, which can be used for a wide range of local activities. Creating a mix-use space in which a local economy can develop around urban shortcuts (mix-street) and urban oasis (small scale green pockets) is using microclimate solutions to facilitate seasonal diversity. The water design is focusing on reusing waste water at a neighborhood scale, converting waste into local economy and energy but also recreating the natural oasis.

宏观规划
Macroscopic Plan

重塑绿洲 Remake the Oasis

城市扩张 City Expansion

连通设计 Design for Connectivity

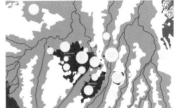
经济平衡 Balanced Economy

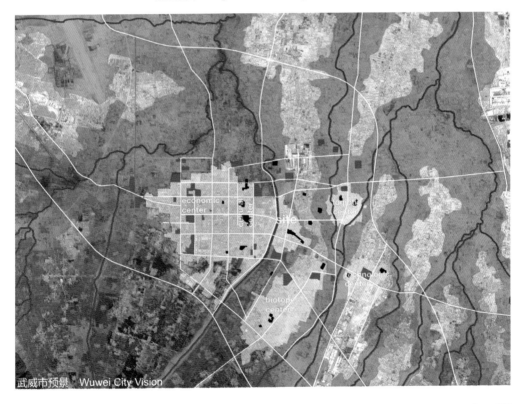
武威市预景 Wuwei City Vision

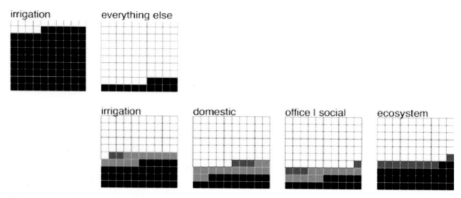

智慧型水利用方案 Designing for Water Smart Solutions

公共空间使用预景 Public Space Use Vision

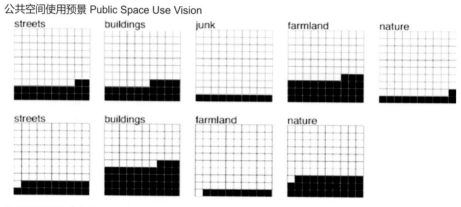

公共空间可达性方案 Designing for Public Space Accessibility

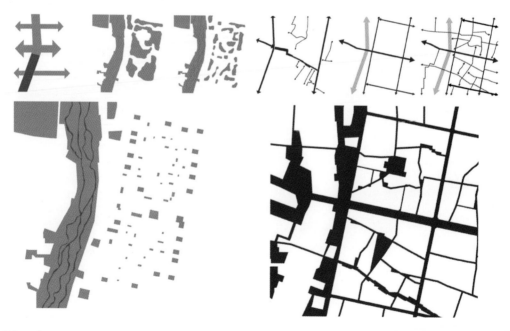

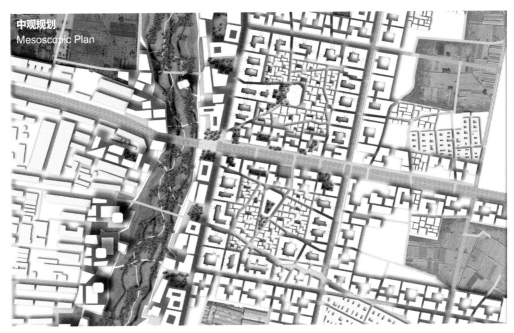

总平面图 Master Plan

开放空间类型
Public Space Typologies

建筑类型 Building Typologies

水景 Water Features

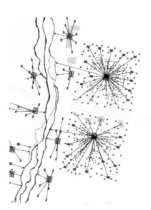

废水利用
Waste Water Use

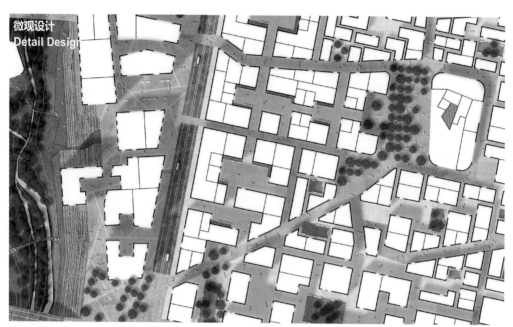

场地平面图 The Site Plan

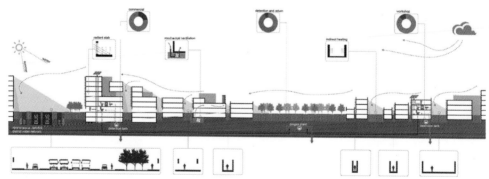

南北剖面图 North-South Diagram

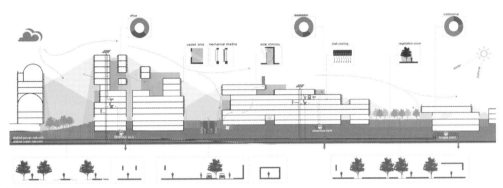

东西剖面图 West-East Diagram

挑战荒漠——城市扩展区可持续城市设计
Challenging Desert——Sustainable Urban Design of City Expansion Area

Cyril Pavlů

武威地处严重干旱地区,气候和水文条件都是未来武威发展的重要因素。本设计的目标是通过合理运用这些限制因素为新的可持续城市环境建立一个框架,这种可持续城市环境能够提供舒适的微气候,便捷的交通以及保留当地的社会价值观。

武威城市扩张的趋势以及新火车站与城区的联系形成了城市的主轴,从西侧的老城区一直延伸到东侧的新火车站。所有的城市核心(CBD、高校、文化机构等)都位于这条主轴的两侧,被高效的公共交通和周边环境所支持。

选址毗邻筹建中的文化中心和石羊河。设计意向是让河道景观中的凉风能够进入到城市肌理中去,并被生长的植物过滤。中等高度并且高密度的带有小院子的平屋顶住房为居民以及周围的公共空间创造了良好的微气候。连续的商业街道以及穿过邻里的公共空间为步行提供了良好的条件,也为公共活动和交通的融合提供了可能。整个设计还包括一个高效水循环系统,由雨水收集,太阳能加热和植被过滤等步骤组成,希望基于当地气候条件为人们创造一个舒适的居住条件。

Wuwei is located in a very challenging arid environment, where the climatic and water conditions are crucial for the future of the city. The goal of this project is to utilize these challenges and use them to create a framework for new sustainable urban environment which should provide a comfortable microclimate, transportation and preserve local social values.

The vision of the city expansion and connection to the newly planned train station

is to create a strong linear city spine, going from the western old city to the new train station in the east. All the city main cores (CBD, university, cultural institutions etc.) are located along this new west-east connection, supported by an efficient public transport and surrounding development.

The site is located adjacent the proposed cultural hub and the river. The proposed development is structured to allow the cool winds from the river valley to penetrate the urban fabric, while being filtered by the lush vegetation that caharcterizes the riverscape. Midrise high density terraced buildings with courtyards provide a pleasant microclimate for residents as well as the public spaces that surround them. The continuous market street and public spaces through the neighborhood provide good walkability and accessibility for public functions and transportation. The development incorporates the rainwater collection, solar warming, and vegetation filtration into a high efficient water-recycling system to provide a sustainable and comfortable living conditions in local climate.

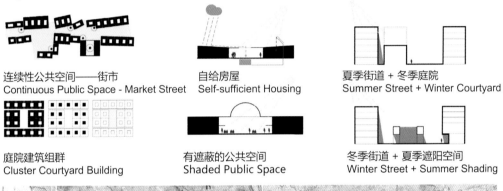

连续性公共空间——街市
Continuous Public Space - Market Street

自给房屋
Self-sufficient Housing

夏季街道 + 冬季庭院
Summer Street + Winter Courtyard

庭院建筑组群
Cluster Courtyard Building

有遮蔽的公共空间
Shaded Public Space

冬季街道 + 夏季遮阳空间
Winter Street + Summer Shading

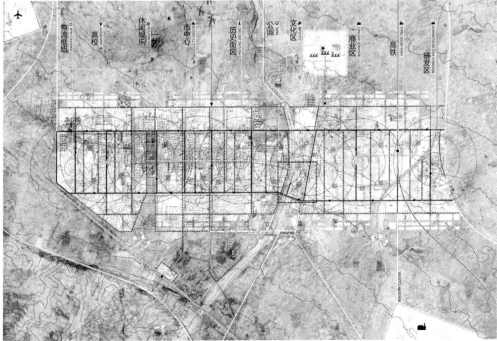

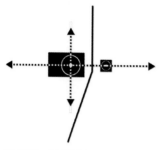

18 世纪—轴线城市

18th Century - City of Axes

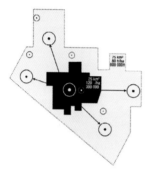

2030 发展规划—多中心扩张

Planned 2030 Development - Polycentric Expansion

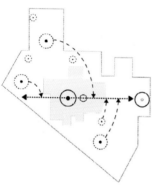

沿未来联系高铁站通道线性组织新的城市中心

Linear Organization of New Urban Cores along the Future Link to Train Station

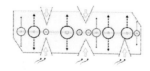

景观渗透城市结构促使通风透气

Urban Structure Penetrated by Landscape to Allow Wind Flow

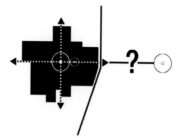

2013—径向扩张，与新火车站缺少联系

2013 - Radial Expansion, Missing Link to New Train Station

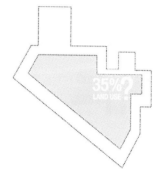

土地利用效率低下导致低密度蔓延

Inefficient Land Use Resulting in Low-dense Sprawl

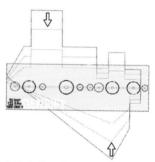

提高密度和增加土地使用多样性—创造更紧凑城市

Densification and Increase of Land Use - Creating More Compact City

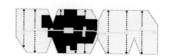

城市主脊—武威的新轴线

The Main Spine of the City - the New Axis of Wuwei

城市主脊
Main City Spine

■ 邻里街市
Neighborhood Market Street

景观 / 气流
Landscape /
Air flow

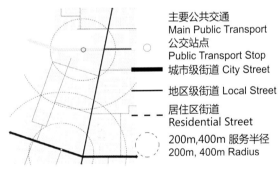

主要公共交通
Main Public Transport
公交站点
Public Transport Stop
—— 城市级街道 City Street
—— 地区级街道 Local Street
--- 居住区街道
Residential Street
○ 200m, 400m 服务半径
200m, 400m Radius

文化节点建筑
Culture Node Buildings
■ 公共服务
Public Services
■ 商业 Commerce
教育 Education
居住 Residential
办公 Office
■ 行政 Administration

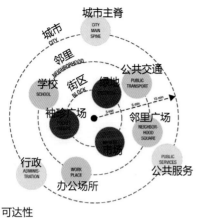

可达性
Accessibility

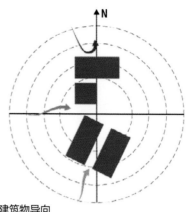

建筑物导向
Building Orientation

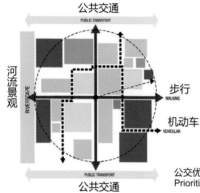

公交优先的非机动车交通
Prioritized Public and Non-vehicular Transportation

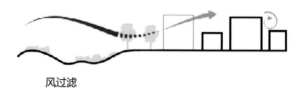

风过滤
Wind Filtration

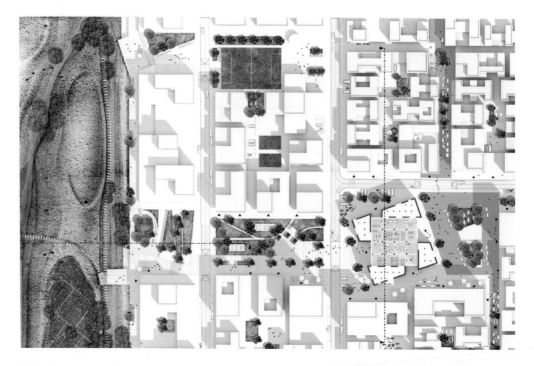

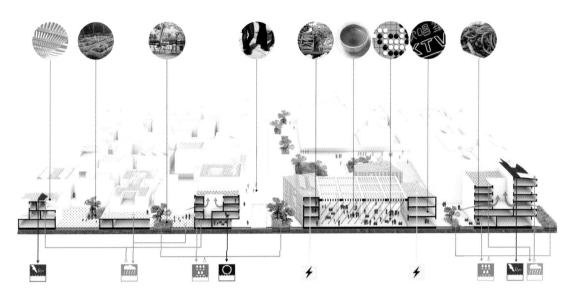
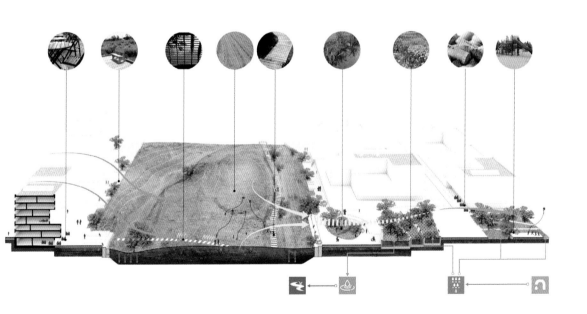

河流之上——城市扩展区可持续城市设计
Over the River—— Sustainable Urban Design of City Expansion Area

Suzanna Rubino

本方案将武威现有城市与其新发展部分相融合,架于改造的河流之上,形成并没有因河流而阻碍的城市。而改造的河流部分作为绿色基础设施,具有补充地下水、休闲、防洪和娱乐等功能。

在城市尺度上,改造后的河道将成为新建城市部分的肌理。在河漫滩廊道部分可以进行高效农业种植,不仅有益于地下水位的回升,还可以为以后城市发展提供空间。通过这种方式,将紧凑的城市建于宜人的景观当中。

本设计拓展了已有的街道和道路结构,创造了丰富的公共空间网络。两条电车轨道线穿过场地,连接了新火车站、城市中心和城市新区。清晰的组团结构形成了狭窄的街区和一些小广场,旨在改善区域的微气候,同时也提供了社交和商业空间以及日常用水的循环系统。不同类型以及保留现有形态的建筑,不仅提供了多样的功能,增加了生活或工作的联系,而且增强了公共街区的活力。同时,拥有多层阶地的河流公园成为密集型发展城市的开敞空间。

The proposal brings the existing city of Wuwei with its new extension together over the restored river. The river park acts as a green infrastructure for water infiltration, treatment, flood control and recreation.

At the city scale, the restored waterways are becoming the framework for the city extension. In this way, a compact city can be built in the oasis landscape. This will allow for the ground water table to refill while efficient agriculture can occur in the river fertile corridors making room for the city development.

The proposed design develops along the existing structure of streets and paths to create a rich network of public spaces. Two tramlines are transect the site, linking it to the new train station, the city centre and the new urbanized areas. The clear block structure defines narrow streets and pocket squares that aim to enhance the microclimate

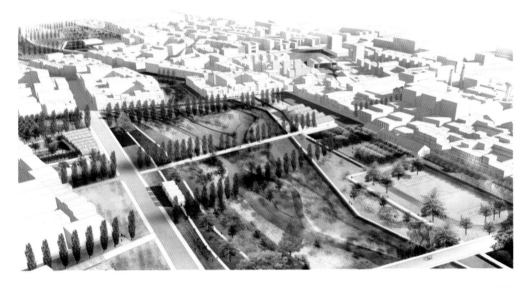

all year round. They also provide social and commercial spaces together with systems to recycle domestic sewage. Different building types will allow for a variety of functions and to enhance the live/work relationship but also to animate street life. Finally, the river park with its different terraces is an open place framed by the densely developed city.

随着人类活动的增强，人工地表水系统已经逐渐取代了自然系统。地表水和地下水均被利用在武威灌溉地区，地下水位下降和春天水流量减少了。因此，想要城市、工业和生活用水可以得到保证，我们应该释放水库存水，补充地下水和维护生态环境发展。

As strengthening of human being's activity, the artificial surface water system has replaced the natural system gradually. As surface water has been imported for irrigation & groundwater exploited in the Wuwei irrigation area, the groundwater level has declined & the spring flux has reduced. Hence, when the water use of city, industry & living could be ensured, we should release water from reservoirs to recharge groundwater & to maintain the eco-environmental development.

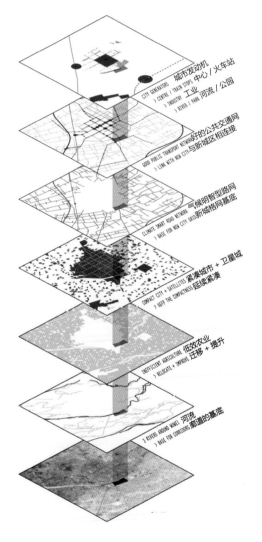

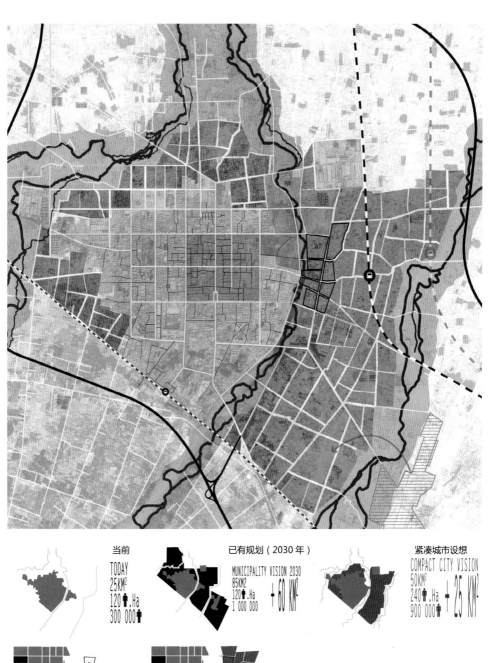
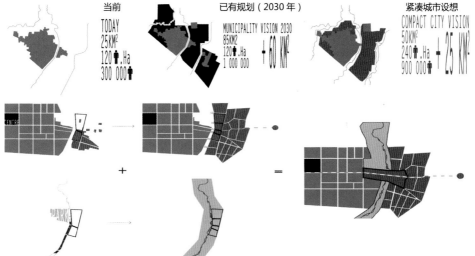

主要的街道结构和公共交通
Main Street Structure and Pubic Transport

小型公共空间网络
Small Public Space Network

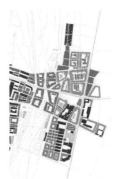
建筑形态
Building Typologies

积极的一层空间和混合功能建筑
Active Ground Floors and mixed Buildings

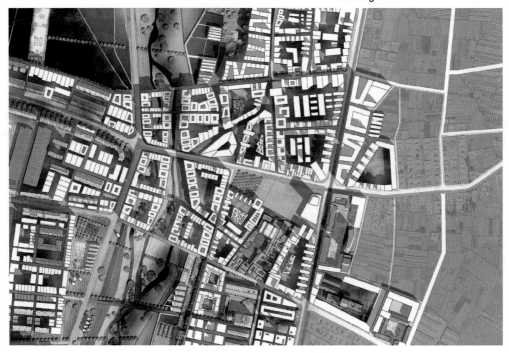

1 - AROUND THE EXISTING
Reuse existing factory buildings to house current site inhabitants
Densify the existing city structure
River park

2 - THE BACK BONE
Tramway streets
Main square and surrounding blocks around administrative buildings

3 - THE BRIDGE
Connect old and new by enlarge the existing bridge as urban fabric

4 - THE RIVER FRONT
Framing the river park with buildings

5 - ...
Start reaching the train station area 2 km further

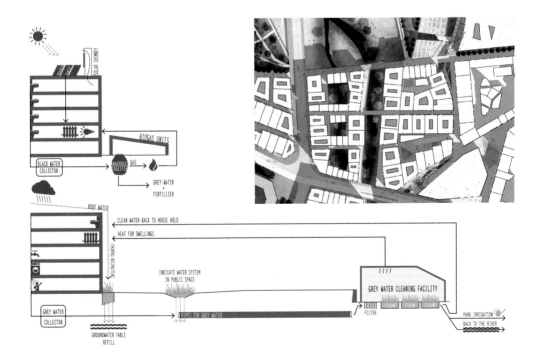

包容的武威——城市扩展区可持续城市设计
Inclusive Wuwei —— Sustainable Urban Design of City Expansion Area

Kateřina Vondrová

项目旨在为武威城市发展提供一种新的方式。到2030年，武威城市人口将翻倍。项目建议沿着主要公共交通环路建立一个具有透水表面的紧凑型城市，各地区中心由公共交通连接。这样的城市结构能够让城市建筑和开放空间产生多种形式的互动，并在整个城市中创造多样化的机遇空间。

设计场地紧邻河流，河床经过修复呈现更多的自然景观，为野生动物提供栖息地，成为一条生物多样性走廊，同时为当地居民提供休闲娱乐的场所。密度较高的城市建筑沿着主要邻里街道分布，且一层为商铺。而低层建筑则用来安置低收入人群。插入建筑之间的公共空间提供多种功能——高密度城市建筑中的综合性公园、社区农田和滨水广场等。最大的广场位于开放空间和主要街道的交接处，是整个设计场地的入口。

The aim of the project is to propose an alternative solution of the future development of Wuwei. The city that is supposed to double its population by 2030 is formed as a permeable but compact urban structure along the main public transport loop that connects the central points of the different districts. This configuration allows different forms of interaction between urban fabric and open landscape and creates diverse edge situations within the whole city.

The design site is located in a close connection to the river. The restored riverbed with more natural conditions provides a habitat for the wildlife and biodiversity corridor but also acts as an important lush green space for the inhabitants. The denser urban structure is located along the main neighborhood streets with commercial units on the ground floor whereas the low-rise buildings in the areas are designed to accommodate the low-income groups. The open space penetrating the fabric provides diverse functions – from formal parks in the dense urban structure to the community farming fields and open riverfront plaza. The main plaza is located on the intersection of the open space stretch and main street and serves as a gate to the area.

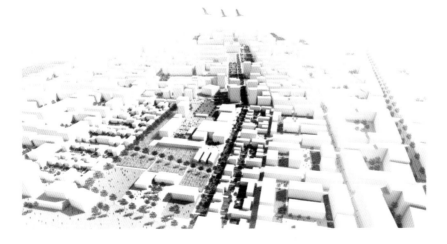

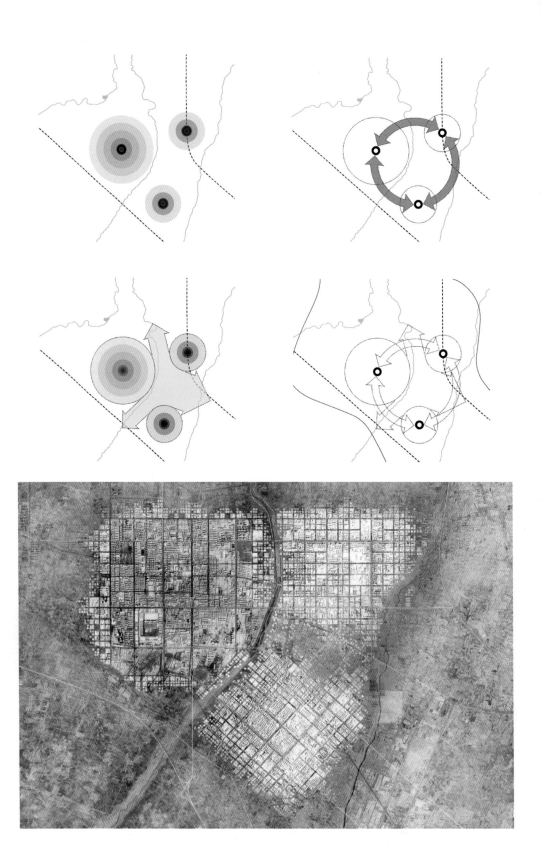

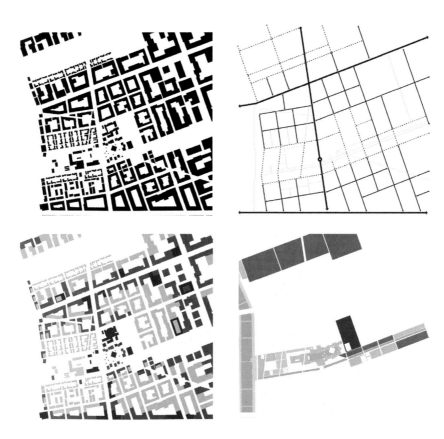

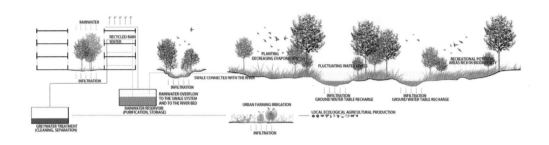

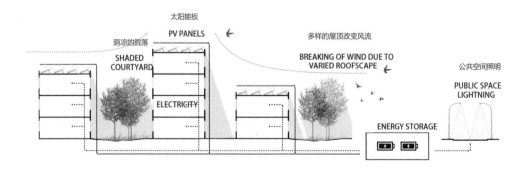

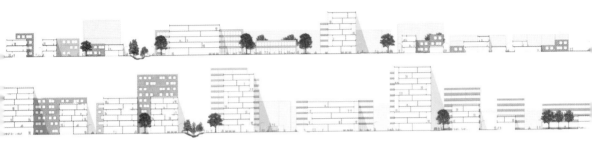

城市群岛——城市扩展区可持续城市设计
The Urban Archipelago——Sustainable Urban Design of City Expansion Area

Sybille de Cussy

城市群岛概念为当前的城市发展提供了一个替代方案。基于当地的气候和区域性缺水的自然条件，如果武威市计划由河流的右岸向规划的火车站发展，城市就必须找到一个新的水管理方案。

该设计将景观作为城市蔓延的限制条件，选择恢复河道自然景观，并将农田迁移到更加肥沃的土地之上。

基于场地上现存的灌溉水渠，将规矩的方格网城市肌理与隐蔽的水渠景观相融合，这些水渠里的溪流将为新设计的城市片区带来更加舒适的微气候，并且制造出风格各异的公共空间，进一步塑造城市肌理。

The urban archipelago proposes an alternative to the current development trend in the city. Considering the dry climate and the lack of water in the region, if Wuwei wants to expand to the right bank of the river toward the planned train station, it will have to deal with a new water management.

The design statement is to use the landscape as the limit of the new urbanization, restoring the riverways and relocating the agriculture in the most fertile soils.

The design proposal suggests to adapt the rigid city grid to the hidden landscape qualities, based on the existing irrigation canals. The canals will allow to stream water to the city for a better microclimate, generating different types of public space, shaping the urban fabric.

The inhabited river park: the riverways are restored, creating meanders and islands. The wide riverbed welcomes the city for an urban continuity between the old and the new city.

The linear promenade: the secondary canals stream the water in the city and generate a vivid commercial and local community life, creating a continuity of public space in the city.

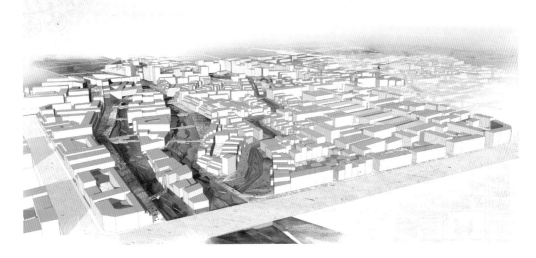

Lack of Water：缺水

高耗水：85%的水用于灌溉
A High Water Consumption:

85%
of the overall available water is used for irrigation

低效的农业：82%的土地是农业用地
An Unefficient Agriculture:

82%
of the land around Wuwei is farmland

未来的城市发展：至2030年人口翻至三倍
A Future Urban Development

3X
the population by 2030

The Landscape Sets the limit for the City
景观作为城市蔓延的限制

关于滨河公园：河道进行恢复，建设蜿蜒的散步道和岛屿。宽阔的河床很好地衔接了老城区和新城区。

关于线性的散步道：次一级的溪流水渠将水带入城市，形成了有活力的商业和社区生活，也提高公共空间的连接度。

关于内向型院落空间：第三级水渠成为宜人的半公共场所，为街区及邻里带来了安静的环境氛围。

The inner courtyards: the tertiary canals are intimate semi-public places, offering a calm atmosphere for the block and the local neighborhood.

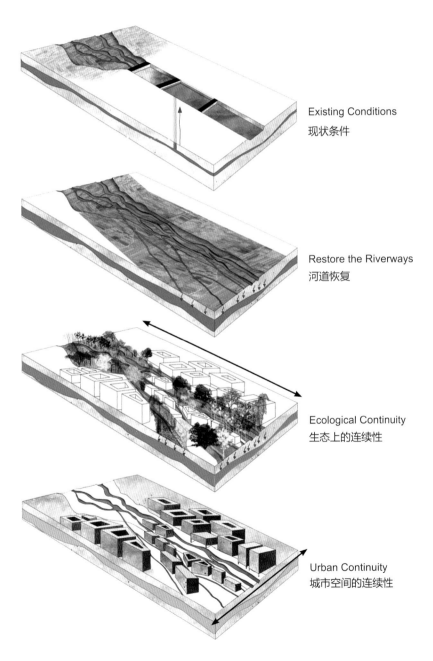

Existing Conditions
现状条件

Restore the Riverways
河道恢复

Ecological Continuity
生态上的连续性

Urban Continuity
城市空间的连续性

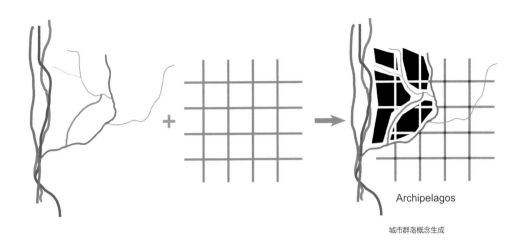

城市群岛概念生成

- Cool Down the City 给城市降温
- Collect and Clean Grey Water 收集和净化污水
- Support Public Space 支持公共空间

概念来源于灌溉水渠
Inspired by the "Acequias" System

场地地形
Topography on Site

最终流向河流的回形水渠系统
Loop System Giving Back to the Rivers

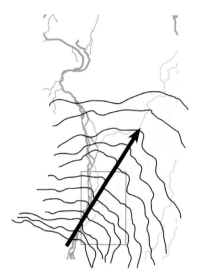
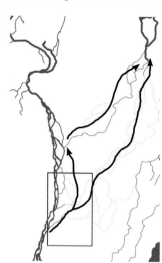
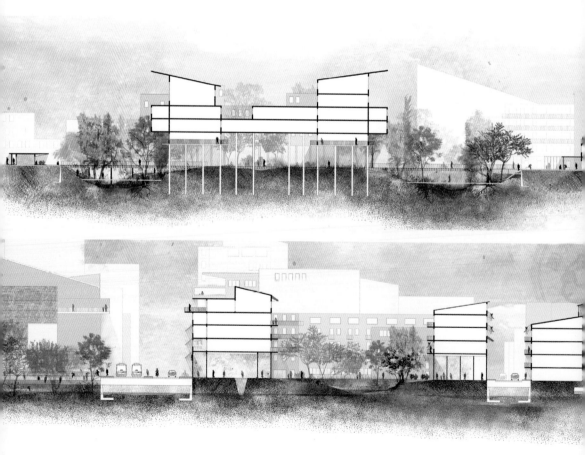

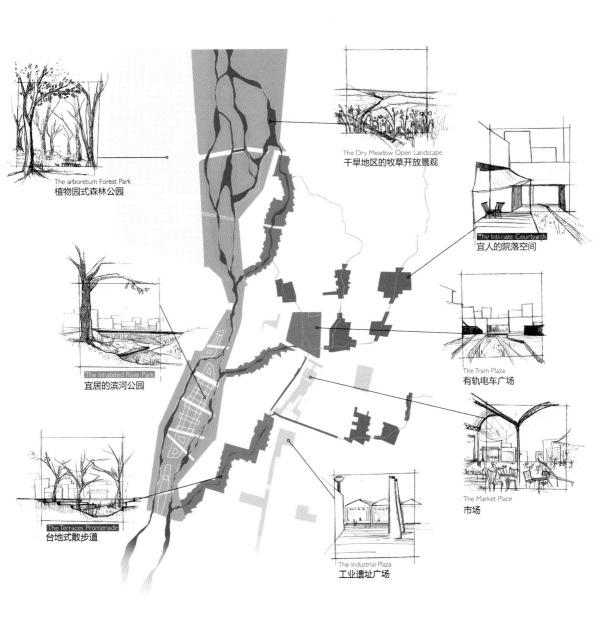

现场踏勘
Site Visit

致谢

特别感谢北京大学基金会设立的"建筑与景观设计学院发展基金"对本书的出版给予的资金支持，该基金由西部发展控股有限公司董事长李西平先生捐赠，他的支持让这本书得以顺利出版。同时，由衷地感谢瑞典隆德大学埃克松约翰逊基金会，他们对于隆德大学建筑学院可持续城市设计专业的师生国际旅行的资金支持使得本次国际研讨课得以开展。

韩西丽 彼得·斯约斯特洛姆 马丁·阿尔福尔克

Acknowledgement

Profound thanks to the "College of Architecture & Landscape Architecture Development Fund", Peking University which provided the financial support for the publication of this book. This fund was donated by Mr. Xiping Li, the chairman of The Western Region Development Holdings co., LTD. His support made a significant contribution to publishing this book. Profound thanks are also extended to the Ax:son Johnson Institute for Sustainable Urban Design, Lund University which provided the financial support for the international travel of students and teachers from Lund University that made this joint workshop possible.

Han Xili Peter Siöström Martin Arfalk